DIGESTIVE SYSTEM MALIGNANCIES

DIGESTIVE SYSTEM MALIGNANCIES

CONSTANTINE P. SPANOS
Associate Professor, Aristotle University of Thessaloniki,
Thessaloniki, Greece

ELSEVIER

ACADEMIC PRESS
An imprint of Elsevier

Academic Press is an imprint of Elsevier
125 London Wall, London EC2Y 5AS, United Kingdom
525 B Street, Suite 1650, San Diego, CA 92101, United States
50 Hampshire Street, 5th Floor, Cambridge, MA 02139, United States
The Boulevard, Langford Lane, Kidlington, Oxford OX5 1GB, United Kingdom

Library of Congress Cataloging-in-Publication Data
A catalog record for this book is available from the Library of Congress

British Library Cataloguing-in-Publication Data
A catalogue record for this book is available from the British Library

ISBN 978-0-323-98369-3

For information on all Academic Press publications
visit our website at https://www.elsevier.com/books-and-journals

Publisher: Stacy Masucci
Editorial Project Manager: Timothy Bennett
Production Project Manager: Maria Bernard
Cover Designer: Miles Hitchen

Typeset by STRAIVE, India

Dedication

To my wife Evangelia

And my daughters Marianna, Fotini, and Athena

Contents

Preface

There are many excellent and comprehensive textbooks that refer to the malignancies of the digestive system. These books originate from a variety of specialties, including surgery, gastroenterology, and oncology. They remain the backbone and primary source of detailed acquisition of knowledge. Although invaluable, these reference books may not be suited for a quick review of material, especially during the early stages of medical knowledge acquisition, that is, in medical school.

Digestive System Malignancies serves as a quick, "at-a-glance" overview of malignant disorders involving the entire digestive tract. A unique feature of this book is the use of infographics. Information on the epidemiology, etiopathogenesis, symptomatology, diagnosis, and treatment are provided using these infographics, which comprise one or two infographic illustrations per topic. The illustrations are supported by brief text, with pertinent, up-to-date bullet points regarding each disorder.

The primary target groups of this book are medical students who are on rotation in surgery or internal medicine, where they encounter patients with digestive malignancies. The book is best used after having studied the material with a standard textbook. *Digestive System Malignancies* is not intended to replace but rather to complement these textbooks. It may also serve as a guide to clinical instructors, when preparing to teach the subjects contained within the book, especially on clinical rounds.

Producing this book has been quite a pleasure. In addition, I have personally gained a bit of knowledge, especially about disorders that are rare. Finally, one realizes the multidisciplinary nature of treating digestive systems malignancies, with all providers belonging to a variety of specialties (medical and radiation oncology, radiology, surgery, nuclear medicine, and pathology) putting in their best efforts to treat their patients. I hope you will enjoy reading this book as much as I enjoyed working on it.

Constantine P. Spanos
1st Department of Surgery, Aristotelian University, Thessaloniki, Greece

Acknowledgments

I am deeply indebted to my wife Evangelia Kokkini, for her constant support and encouragement throughout my professional career, despite being a very busy teacher and mother to our three girls. Furthermore, I am most grateful to my professors and mentors who played an important role in my surgical training—William C. Mackey (Tufts University) and Robert W. Beart, Jr. (University of Southern California). Andreas M. Kaiser (City of Hope), a mentor and a true friend, has supported me in every publishing effort I have undertaken. He provided invaluable insight and encouragement. My father, a cardiac surgeon, who inspired me to go into the surgical field, deserves an enormous amount of credit; he has been a big fan of the infographic format of this book. Finally, a great deal of appreciation goes to my medical students and residents, for whom this book was written. They really do push me to stay current and sharp in the ever-changing world of our profession.

CHAPTER 1

Esophageal cancer

- Esophageal cancer is the eighth commonest solid organ tumor worldwide and the sixth most common cause of cancer death. It is endemic in Central Asia, India, and China.
- There are two main histologic types of esophageal cancer: squamous cell cancer (SCC) and adenocarcinoma (AC). Globally, SCC is the most common type; the incidence of adenocarcinoma has surpassed SCC in Western countries.
- Etiologic factors associated with SCC include:
  - Smoking
  - Alcohol abuse
  - Nutritional deficiencies
  - Environmental factors
  - Achalasia
  - HPV
  - PlummerVinson syndrome
  - Radiation
  - Caustic ingestion
- Etiologic factors associated with AC include:
  - Advanced age
  - Gastroesophageal reflux disease (GERD)
  - Barrett's esophagus/intestinal metaplasia
  - Male sex
  - Obesity
- When esophageal cancer presents with clinical symptoms, it is usually at an advanced stage. Esophageal cancer is locally advanced, metastatic or incurable in 50%–80% of cases on presentation.
- Dysphagia occurs when 2/3 of the lumen is obstructed. Dysphagia begins with solid food and progresses to intake of fluids.
- Hoarseness occurs when the recurrent laryngeal nerve is involved.
- Anorexia and weight loss are common.

Digestive System Malignancies
https://doi.org/10.1016/B978-0-323-98369-3.00010-1

- Other symptoms include odynophagia, long-standing GERD, hiccups (involvement of phrenic nerve) and rarely hematemesis.
- Early diagnosis can be made with endoscopic screening (prevalent in Asia), as well as during surveillance for Barrett's esophagus.
- Initial diagnosis of esophageal cancer is made by endoscopy (esophagogastroduodenoscopy-EGD). This will determine the location and extent of tumor, allow for biopsies. Chromoendoscopy and narrow-band imaging may facilitate the diagnosis.
- In several centers, EGD is preceded by an upper gastrointestinal contrast swallow study. This may also depict the location of the tumor, degree of obstruction and help to plan the EGD.
- Pretreatment staging is an essential step in the management of esophageal cancer. Therapy is based on pretreatment staging assessment. This is performed using several modalities:
    - Computed tomography (CT) detects distant metastases. It is less sensitive/specific for early T stage disease. It is also not accurate in the assessment of lymph node metastases.
    - FDG-PET (positron emission tomography) can be used for the detection of metastatic disease.
    - Endoscopic ultrasound (EUS) can assess the tumor depth invasion, depict suspicious lymph nodes (large, hypoechoic with rounded appearance), and allow for biopsy. Lymph node biopsy through the esophageal tumor is contraindicated due to risk of seeding. In addition, invasion of pleura, bronchi, lung, heart, and spine can be assessed.
    - Endoscopic mucosal resection (EMR) can be used as a diagnostic and staging modality. It provides an adequate tissue sample, accurate depth of invasion assessment, and may lead to resection of the entire lesion, when it is <2 cm in size. EMR has therapeutic potential in early-stage small lesions.
    - Laparoscopy is routinely used as a staging modality in some centers. It allows for the assessment of the gastroesophageal (GE) junction, the liver (for metastases), the peritoneum (for peritoneal surface malignancy), and the celiac axis (for node biopsy). A feeding jejunostomy can also be placed laparoscopically.
- Treatment incudes esophagectomy, endoscopic resection, and esophageal-sparing modalities (endoscopic mucosal ablation, radiofrequency ablation, photodynamic therapy, and cryotherapy).

- Therapy is based on pretreatment staging assessment.
  - Endoscopic resection can be performed in patients with small tumors (<2 cm)involving less than one-third of the esophageal circumference and depth of invasion limited to the mucosa. Complications include perforation, bleeding, and stricture formation.
  - **TisN0Mo** lesions can be treated with endoscopic mucosal ablation or esophagectomy. Esophagectomy is associated with a 90% 5-year survival rate, but with an operative mortality of 3%–12% and 30%–50% morbidity rate. Intense endoscopic surveillance is an option in highly selected patients.
  - **T1aN0M0** lesions can be treated with esophageal sparing modalities or esophagectomy; lymph node metastases occur in less than 2% of patients.
  - **T1bN0M0** lesions can be treated with esophageal sparing modalities if less than the proximal 1/3 of the submucosa (sm1) has been invaded by tumor. If the submucosal invasion is deeper (sm2, sm3), esophagectomy with regional lymphadenectomy is recommended.
  - **T2-4N0-3M0** lesions are usually treated with neoadjuvant chemoradiotherapy. This may increase the ability to achieve an R0 resection. The CROSS trial has been influential with regards to this approach; median survival in patients treated with neoadjuvant chemoradiotherapy followed by surgery was 49 months, compared with 24 months in patients treated with surgery alone.
- Surgical procedures used in the treatment of esophageal cancer include open surgery, minimally invasive, or robotically assisted procedures. The most common operations performed are the following:
  - **The Ivor-Lewis procedure**. This consists of a laparotomy/laparoscopy and right thoracotomy/thoracoscopy. It is used for tumors of the lower and middle third of the esophagus. During this operation the azygous vein is divided, the esophagus resected, and a conduit (gastric or colonic) is constructed to establish alimentary tract continuity.
  - **The McKeown procedure**. This consists of a right posterolateral thoracotomy (dissection of esophagus/lymph nodes), laparotomy (gastric conduit/colon interposition conduit construction), and a left oblique neck incision (anastomosis of esophagus with conduit). It can also be performed with minimally invasive/robotically

assisted techniques. It is ideal for tumors of the middle/upper third of the esophagus. The cervical anastomosis performed mitigates the mediastinitis associated with anastomotic leaks in the Ivor-Lewis procedure.

- **Trans hiatal esophagectomy**. A midline laparotomy is used for full gastric mobilization; a diaphragmatic incision is made, and blunt esophageal dissection/mobilization is undertaken. A left oblique neck incision provides access for proximal esophageal mobilization and the esophageal/conduit anastomosis. It can also be performed with minimally invasive/robotically assisted techniques. It is preferred in patients who are unable to tolerate pulmonary/thoracic morbidity associated with the Ivor-Lewis and McKeown procedures.
- **Left thoraco-abdominal esophagectomy.** This is utilized in patients with large, bulky tumors located in the vicinity of the esophageal hiatus.

- Esophagectomies are associated with significant morbidity. Of note, respiratory complications occur in 50%–65% of cases. Preoperative optimization as well as the use of epidural catheters/intercostal anesthetic blockade is widely used. Significant surgical complications include anastomotic leak and stricture, chyle leak, and bleeding.
- Adjuvant chemoradiotherapy is recommended in patients with high risks of local/systemic recurrence (T3, T4, N disease, and R1/R2 resections).
- Palliative treatment for advanced/unresectable cancers or tumors in patients who are unable to tolerate esophagectomy include chemotherapy, chemoradiation, stents, balloon dilatation, and argon-plasma coagulation.
- **The Siewert classification** is used for tumors in the vicinity of the EG junction. They encompass all tumors with an epicenter within 5 cm proximal/distal to the EG junction; all are adenocarcinomas.
  - **Siewert type I lesions**: 1–5 cm proximal to EG junction. Usually arise from Barrett's esophagus.
  - **Siewert type II lesions**: 1 cm proximal–2 cm distal to EG junction. Usually arise from the epithelium of the gastric cardia.
  - **Siewert type III lesions**: 2–5 cm distal to EG junction. This is a subcardial gastric adenocarcinoma.

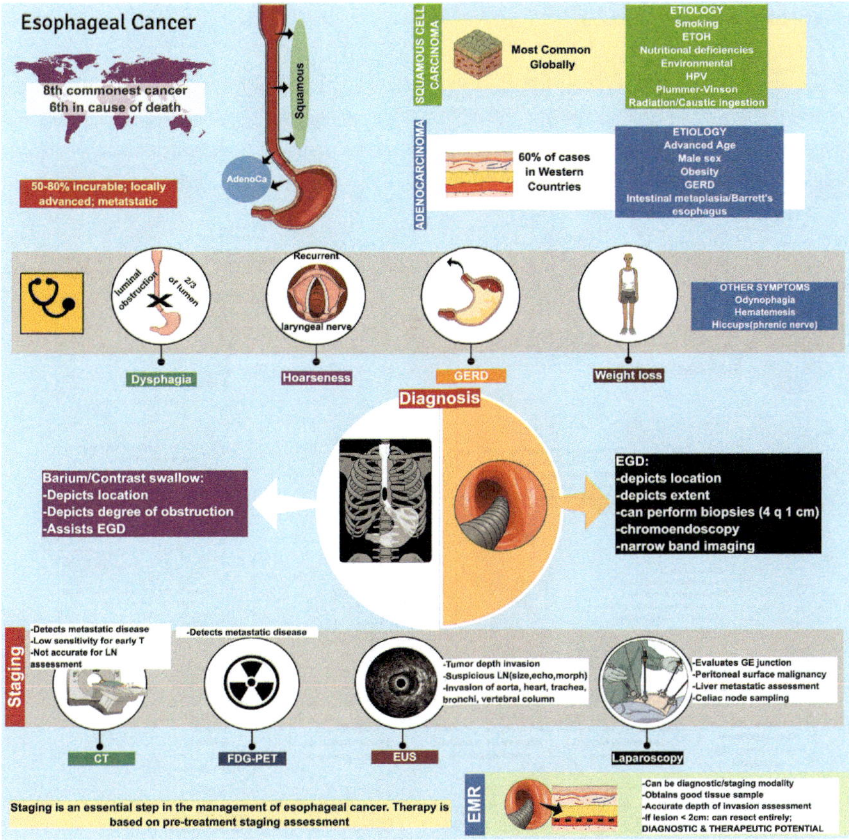

Fig. 1 Esophageal cancer: Epidemiology and evaluation.

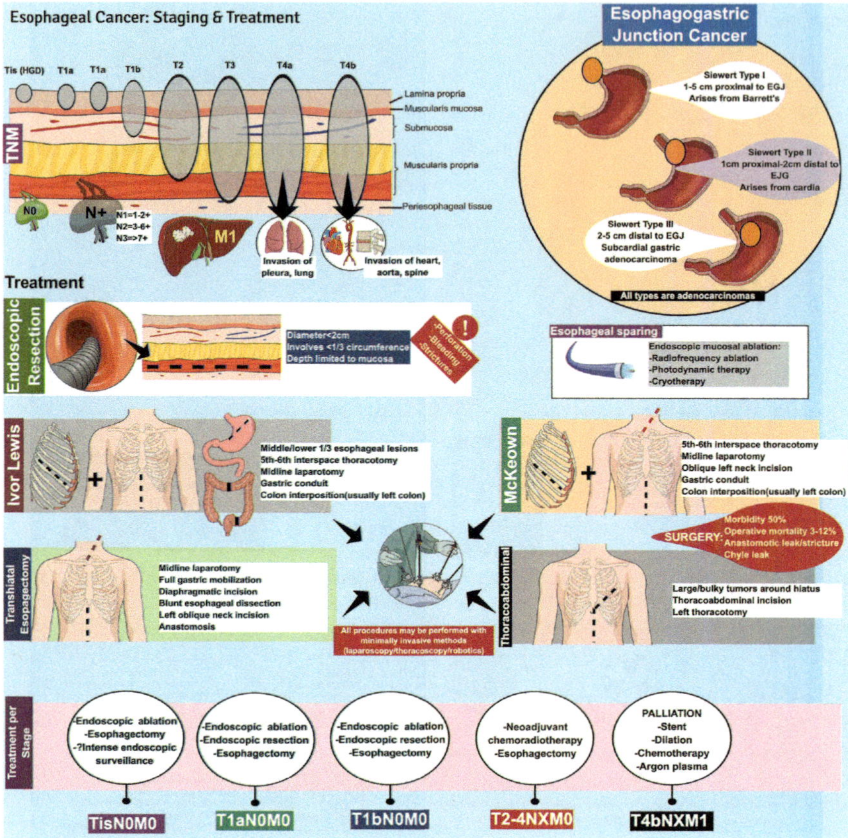

Fig. 2 Esophageal cancer: Staging and treatment.

Further reading

Jobe BA, Landreneau RJ, Habib F. The management of esophageal cancer. In: Cameron JL, Cameron AM, editors. Current surgical therapy. 12th ed. Philadelphia: Elsevier; 2017. p. 53–62.

Kohtz PD, McCarter MD, Weyant M. Esophageal cancer. In: RC MI, Schulick RD, editors. Surgical decision making. 6th ed. Philadelphia, PA: Elsevier; 2020. p. 236–8.

Miyata H, Yamasaki M, Kurokawa Y, et al. Multimodal treatment for resectable esophageal cancer. Gen Thorac Cardiovasc Surg 2011;59:461.

Kelly RJ, Ajani JA, Kuzdal J, et al. Adjuvant nivolumab in resected esophageal or gastroesophageal junction cancer. N Engl J Med 2021;384:1191–203.

CHAPTER 2

Gastric cancer

- Gastric cancer is the fifth most common cancer, and the third commonest cause of cancer-related death. Most cases occur in China, Japan, Korea, Central, and South America. In North America, Australia, and North Africa, the incidence is not as high. Overall, the incidence of gastric cancer has been decreasing. In the United States, 24,600 cases and 10,700 deaths were due to gastric cancer (2015 data); a higher incidence was noted in African Americans, Hispanics, and Native Americans.
- In Western countries there has been an increase in proximally located gastric cancer. In Japan, distally located tumors predominate.
- Risk factors for the development of gastric cancer include *Helicobacter pylori* infection (associated with distal gastric cancer), gastric polyps, chronic exposure to nitrosamines, smoking, pernicious anemia, and previous gastric surgery.
- In 15% of cases there is a family history of gastric cancer. Inherited gastric cancer syndromes include hereditary diffuse gastric cancer (E-cadherin/CDH-1 mutation), familial adenomatous polyposis (FAP), and Lynch syndrome (HNPCC).
- Most gastric cancers are adenocarcinomas. There are two histologic subtypes (Lauren classification):
    - **Intestinal type**: well-differentiated, forms in gastric glands, occurs in older patients, is associated with *H. pylori* infection and distal lesions. It is more prevalent in China, Korea, and Japan.
    - **Diffuse type**: poorly differentiated, E-cadherin mutation frequent, with loss of cellular adhesion. It is associated with linitis plastica, proximal lesions, and has a worse prognosis. It is more frequent in Western populations.
- Symptomatology is associated with advanced lesions. Patients may complain of abdominal pain; this is vague and mild initially, progressing to severe pain. Anorexia, decreased caloric intake, and weight loss are common. Proximal cancers may present with dysphagia; pseudo-achalasia may result from tumor invading the myenteric Auerbach plexus. Gastric outlet obstruction in large tumors, distal tumors or linitis plastica present with nausea and vomiting.

Digestive System Malignancies
https://doi.org/10.1016/B978-0-323-98369-3.00001-0

- Occult blood loss from the tumor results in anemia; overt bleeding may present with melena or hematemesis.
- Advanced and/or metastatic disease may present with:
  - Palpable eponymous nodes (Sister Mary Joseph or periumbilical nodes; Virchow's or left supraclavicular nodes)
  - Palpable pelvic peritoneal surface disease (Blumer's shelf; palpable on digital rectal exam)
  - Palpable ovarian tumors (Krukenberg tumors)
  - Ascites
  - Palpable abdominal mass (peritoneal surface malignancy; hepatic metastases)
  - Jaundice
- Paraneoplastic syndromes associated with gastric cancer include:
  - Diffuse seborrheic keratosis (Leser-Trelat syndrome)
  - Acanthosis nigricans
  - Membranous nephropathy
  - Microangiopathic hemolytic anemia
  - Hypercoagulability
- Diagnostic confirmation is usually made with esophagogastroduodenoscopy (EGD) with biopsies.
- Pretreatment staging is important for the overall treatment plan. Modalities that may facilitate staging include CT scan, endoscopic ultrasound (EUS), FDG/PET, and pretreatment laparoscopy.
- CT scan of the chest, abdomen, and pelvis is useful to detect distant metastatic disease. It is not as sensitive or specific in detecting peritoneal surface malignancy (carcinomatosis). Peritoneal carcinomatosis may be present in 20%–30% of negative scans.
- EUS is useful in assessing the depth of tumor invasion (T stage), as well as possible nodal involvement (N stage). It is particularly helpful in ruling out submucosal invasion, which may allow for endoscopic resection in selected patients.
- FDG/PET may be performed in patients with T2 and greater lesions, to detect distant metastatic disease not depicted on CT scan. It is not as sensitive or specific in detecting peritoneal surface malignancy (carcinomatosis).
- In several institutions, pretreatment laparoscopy is routine. Benefits include increased sensitivity in detecting distant metastatic disease, peritoneal surface malignancy, and the ability to perform peritoneal fluid

cytology. Positive peritoneal "washings" portend poor prognosis. A feeding jejunostomy may also be placed for pretreatment nutritional optimization at the time of laparoscopy.

- Early gastric cancer is defined as T1 (invasion into submucosa)-any N. Treatment may lead to a 90% 5-year survival.
- Small (<2 cm without ulceration), well-differentiated lesions confined to the mucosa without lymphovascular invasion may be treated with endoscopic submucosal dissection (ESD) or endoscopic mucosal resection (EMR) in selected patients.
- Gastrectomy is the mainstay of treatment of gastric cancer. The main objective is R0 resection, with gross margins >5 cm and negative frozen-section margins. Distal lesions may be treated with a subtotal gastrectomy, more proximal lesions with total gastrectomy.
- Emphasis has been placed on the extent of lympadenectomy during gastrectomy. Lymph node dissections are classified anatomically as:
    - **D1**: during which perigastric nodes are resected only.
    - **D2**: during which nodes of the celiac axis, and splenic hilum are resected in addition.
    - **D3**: during which para-aortic and nodes in the porta hepatis are resected in addition to D1/D2 nodes.
- In most centers with a high volume of gastric cancer, D2 dissection is performed. This is based on metanalyses that demonstrate decreased cancer-specific mortality with this operation.
- Alimentary tract continuity is established with a Billroth II procedure or a Roux-Y reconstruction. An omentectomy is always performed.
- Combined modality therapy is considered in patiens whom have tumors that are T2 or greater. **Perioperative chemotherapy** is used frequently; this entails three cycles of preoperative chemotherapy followed by surgery, followed by three additional cycles of chemotherapy. The British MAGIC trial demonstrated a 25% reduction in cancer-specific mortality in the group treated with perioperative therapy when compared with patients treated with surgery alone.
- Adjuvant (postoperative) chemotherapy may be offered to patients with nodal metastases; this may confer a slight disease-free and/or overall survival compared with D2 dissection alone (CLASSIC trial).
- Adjuvant chemoradiation therapy has been used in patients who undergo gastrectomy with incomplete D2 dissection with a slight benefit in the 3-year disease-free, but not overall survival.

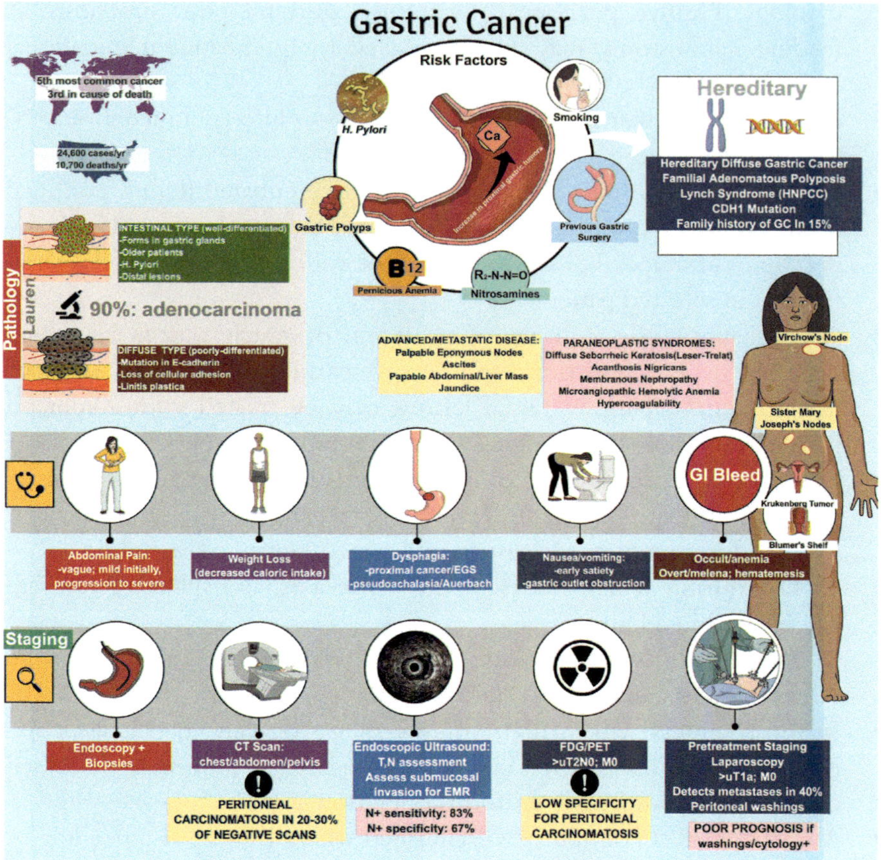

Fig. 1 Gastric cancer: Risk factors, classification, and evaluation.

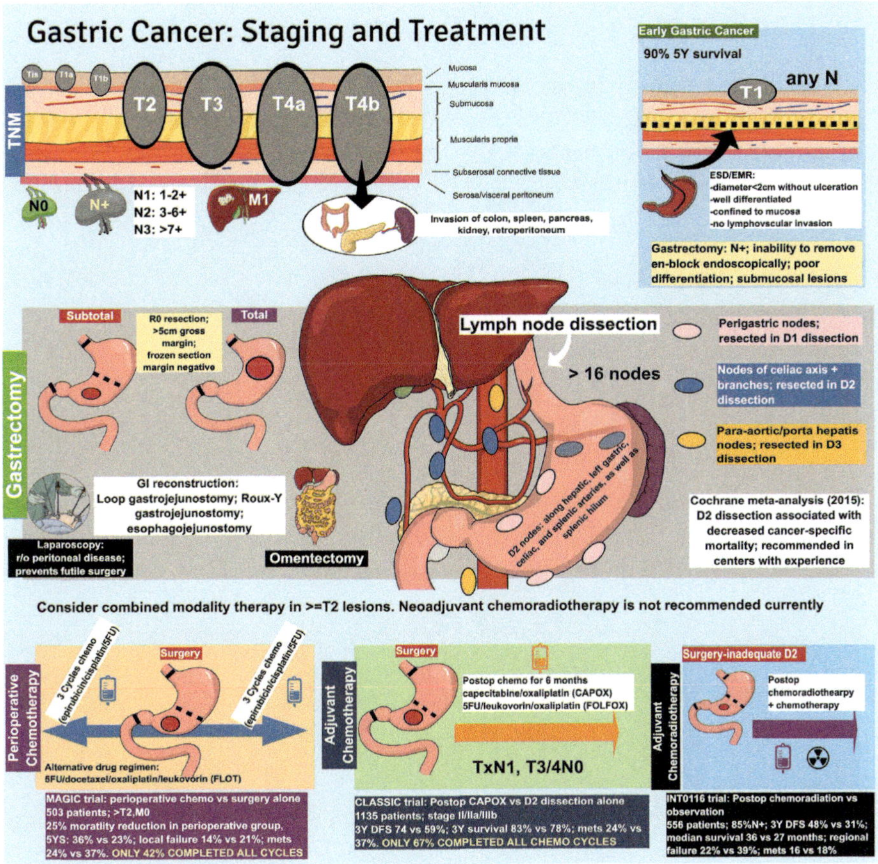

Fig. 2 Gastric cancer: Staging and treatment.

Further reading

[1] McFadden DW, Brenner BM. The management of gastric adenocarcinoma. In: Cameron JL, Cameron AM, editors. Current surgical therapy. 12th ed. Philadelphia, PA: Elsevier; 2017. p. 94–8.

[2] Selby LV, McCarter MD. Gastric cancer. In: McIntyre Jr RC, Schulick RD, editors. Surgical decision making. 6th ed. Philadelphia, PA: Elsevier; 2020. p. 240–1.

[3] Simon M, Mal F, Perniceni T, et al. Accuracy of staging laparoscopy in detecting peritoneal dissemination in patients with gastroesophageal adenocarcinoma. Dis Esophagus 2016;29:236.

CHAPTER 3

Primary gastrointestinal lymphoma

- Primary gastrointestinal lymphomas (PGIL) are rare gastrointestinal malignancies. They comprise 1%–5% of GI tumors. Most are non-Hodgkin lymphomas. Of note, secondary involvement of the GI tract in non-Hodgkin lymphoma patients is more common than PGL.
- Gastric lymphoma is the most common site of PGIL.
- Predisposing factors associated with PGIL include:
  - *Helicobacter pylori* (associated with extranodal marginal zone B-cell lymphoma of mucosa-associated lymphoid tissue-MALT).
  - Celiac disease (Enteropathy-associated T-cell lymphoma).
  - Autoimmune disorders (rheumatoid arthritis, Sjogren's syndrome, Systemic lupus, and granulomatosis + polyangitis).
  - Congenital immunodeficiency (Wiskott-Aldrich, ataxia-telagiectasia, and X-linked agammaglobulinemia).
  - Acquired immunodeficiency (AIDS, iatrogenic/pharmacologic).
- Esophageal lymphoma is rare; it occurs in the distal esophagus.
- Primary gastric lymphoma (PGL) is the most common extranodal site of lymphoma (68%–75% of PGILs, 10% of lymphomas, and 3% of gastric neoplasms). Two main histologic types of PGL are encountered: extranodal marginal zone B-cell lymphoma of mucosa-associated lymphoid tissue-MALT and diffuse large B-cell lymphoma.
- PGL symptoms include epigastric pain, anorexia, early satiety, weight loss, nausea/vomiting, and occult or overt GI bleeding. "B" symptoms, such as night sweats and fever occur in 12% of cases. Lymphadenopathy and/or a palpable mass are signs of advanced disease.
- Diagnostic modalities for PGL include endoscopy with biopsy. Since PGL is usually submucosal, endoscopic ultrasound-guided fine-needle aspiration biopsy (EUS/FNAB) may be necessary. Biopsy specimens should be tested for *H. pylori*. Fluorescent in situ hybridization

Digestive System Malignancies
https://doi.org/10.1016/B978-0-323-98369-3.00011-3

(FISH/PCR) should also be undertaken to detect t(11;18) translocation; when present, patients are nonresponsive to *H. pylori* eradication treatment.

- Small intestinal lymphoma accounts for 13%–30% of PGIL. Main subtypes include EATL, diffuse large B-cell lymphoma, Burkitt's lymphoma, and mantle-cell lymphoma.
- Mantle-cell and Burkitt's lymphoma are associated with young age, low socioeconomic status, parasitic infestation, infantile enteritis, and *Campylobacter jejuni* infection.
- Symptoms include abdominal pain, diarrhea, malabsorption, clubbing, ankle edema, and weight loss. EATL symptoms include GI bleeding, obstruction, perforation, and clinical deterioration of celiac disease.
- Colorectal lymphoma is also rare. The majority are located proximal to the hepatic flexure. The ileocecal area is the most common location of colorectal lymphoma. Mantle cell, Burkitt's, follicular, and diffuse large B-cell lymphoma are the most common histological types.
- The most common presentation is a tender mass, usually in the RLQ. Other symptoms include abdominal pain, anemia, rectal bleeding, and fever. Bowel obstruction occurs in 25% of cases.
- The general diagnostic workup of PGIL may include:
  - A careful physical exam, including eyes, ear, nose, and throat; careful palpation for lymphadenopathy, hepatomegaly/mass, splenomegaly, as well as careful examination for skin lesions.
  - CT scan, which assesses the extent of disease, and involvement of local/distal lymph nodes and other organs.
  - Endoscopy, including EUS ± FNAB. Push enteroscopy is useful in the diagnosis of proximal small bowel tumors; large jejunal ulcers are characteristic in EATL.
  - FDG-PET may be used in the diagnosis/staging of PGIL. Diffuse large B-cell lymphomas are FDG avid; MALT has variable FDG avidity.
- The Lugano staging system is most commonly used.
  - Stage I lesions are confined to the GI tract and are simple and noncontiguous.

- Stage II lesions extend into the abdomen from the primary GI site, with local (II1) or distant (II2) nodal involvement. Stage IIE denotes invasion of adjacent organs.
- Stage IV lesions are characterized by disseminated nodal involvement in addition to the GI lesion, or concomitant supradiaphragmatic nodes.
- There is no Stage III in the Lugano classification.

- Treatment of gastric MALT lymphoma depends on the stage and the presence of *H. pylori*. In patients with Lugano stage I/II, *H. pylori*-positive MALT, *H. pylori* eradication is the initial therapy of choice. This has been associated with a 50%–85% complete pathological response, with a mean time to eradication/response of 5 months to 3 years. Serial endoscopies are required; patients may need up to three eradication regimens.
- Relapse in patients with MALT occurs in 22% and may be treated with radiation therapy or a combination of rituximab ± chemotherapy. The latter may also be used as initial therapy for Lugano stage IV MALT.
- In patients with *H. pylori*-negative MALT or with t(11;18) translocation or persistent disease after other forms of therapy, treatment options include radiation therapy and rituximab ± chemotherapy.
- Gastrectomy as primary treatment is performed for bleeding, perforation, and obstruction.
- Treatment of Intestinal lymphoma for Lugano stage I/II1 includes resection followed by chemotherapy. Lugano stage II2/IIE/IV are treated with either chemotherapy or stem cell transplantation. These lesions are at high risk for perforation. Emergency surgery is performed for bleeding, perforation, and obstruction.
- Colonic lymphoma is treated with resection followed by chemotherapy. Chemotherapy alone portends a high risk of perforation.
- Rectal lymphoma may be treated by radiation, with possible subsequent resection, and adjuvant chemotherapy.
- Five-year survival rates for resection and adjuvant chemotherapy of colorectal lymphoma are 40%–83%. T-cell lymphomas have worse prognosis.

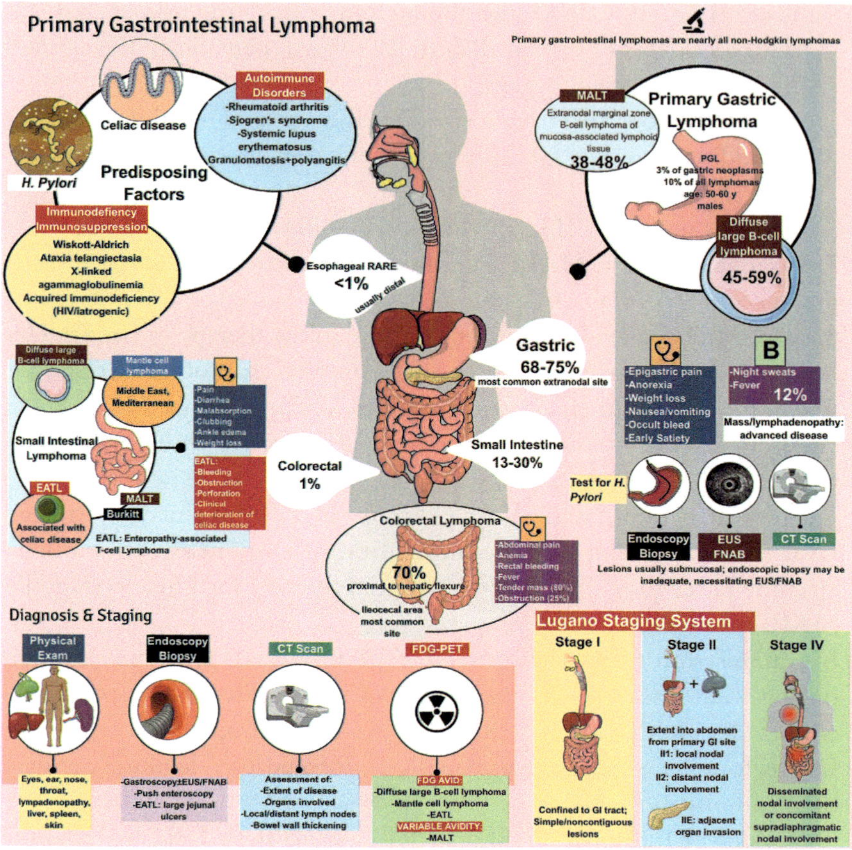

Fig. 1 Primary gastrointestinal lymphoma: Classification, diagnosis, and staging.

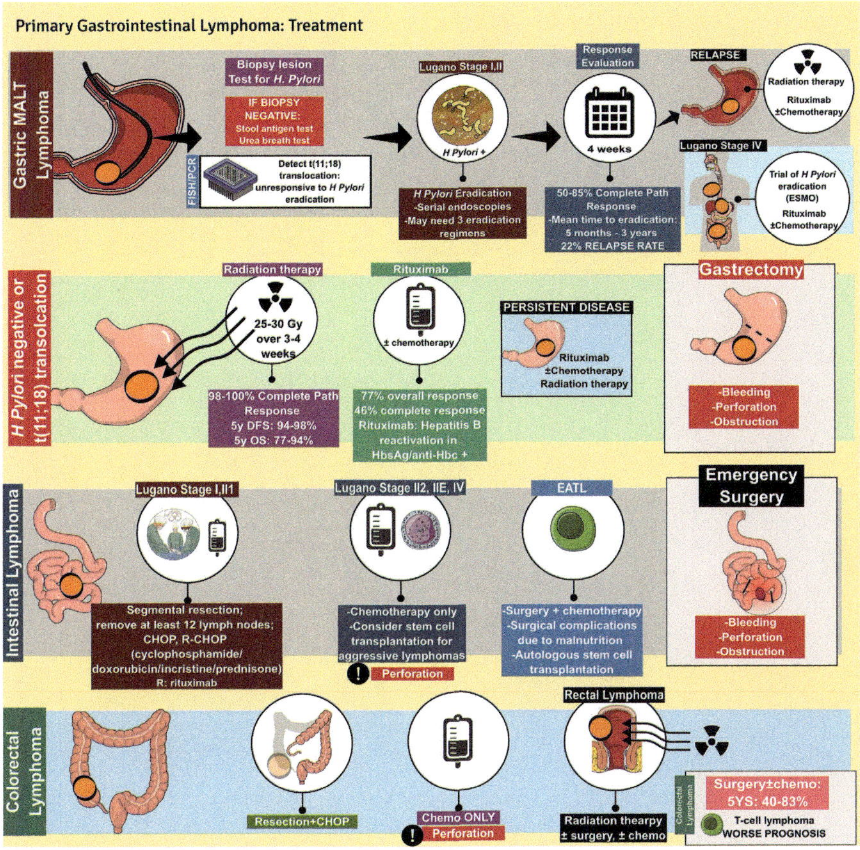

Fig. 2 Primary gastrointestinal lymphoma: Treatment.

Further reading

[1] Chapman B, Gajdos C. GI lymphoma. In: McIntyre Jr RC, Schulick RD, editors. Surgical decision making. 6th ed. Philadelphia, PA: Elsevier; 2020. p. 274.

[2] Zucca E, Arcaini L, Buske C, et al. Marginal zone lymphomas: ESMO clinical practice guidelines for diagnosis, treatment and follow-up. Ann Oncol 2020;31:17.

[3] Rohatiner A, d'Amore F, Coiffier B, et al. Report on a workshop convened to discuss the pathological and staging classifications of gastrointestinal tract lymphoma. Ann Oncol 1994;5:397.

[4] Helmink BA, Bailey CE, Tarpley JL. The management of small bowel tumors. In: Cameron JL, Cameron AM, editors. Current surgical therapy. 12th ed. Philadelphia, PA: Elsevier; 2017. p. 121–7.

CHAPTER 4

Gastrointestinal stromal tumors (GIST)

- Gastrointestinal stromal tumors (GIST) are the most common mesenchymal tumors of the gastrointestinal tract. In fact, they are the most common subtype of sarcoma overall. They were formally termed leiomyosarcomas. They comprise 1% of GI malignancies.
- GISTs arise from the interstitial cells of Cajal in the myenteric nerve plexus of hollow intra-abdominal organs. The tumors develop after activation of tyrosine kinase; an activating mutation in certain proto-oncogenes is required for this.
- Approximately 75% of GISTs bear mutation in c-KIT; 10% have mutations in the PDGFRα gene. "Wild-type" GISTs do not have these mutations. The majority of GISTs express the cKIT (CD117) protein.
- GISTs may be associated with the Carney-Stratakis syndrome (GIST + paraganglioma), Carney's triad (GIST/paraganglioma/pulmonary chordoma), and neurofibromatosis type 1 (GIST/neurofibromas/glioma/malignant peripheral nerve sheath tumors).
- The distribution and frequency of GISTs throughout the GI tract is as follows:
  - Esophagus: < 1%
  - Stomach: 60%
  - Duodenum: 5%
  - Small intestine: 30%
  - Colorectal: 5%
- GISTs have three histologic pattern characteristics: spindle-cell type (most common), epithelioid type, and mixed-type GISTs.
- Risk stratification of GISTs takes into account size (> 5 cm, worse prognosis), the mitotic index (> 5/50 per high-power field, worse prognosis), and location (nongastric GISTS, worse prognosis). Molecular profiling after FNA biopsy may also provide additional prognostic indicators.
- Clinical manifestations include:
  - GI bleeding (overt/occult), in 28%–50% of cases

Digestive System Malignancies
https://doi.org/10.1016/B978-0-323-98369-3.00003-4

    - Abdominal pain
    - Acute abdomen
    - Incidental finding on endoscopy or imaging
- Diagnostic modalities include:
    - Endoscopy: a submucosal mass with central ulceration ± bleeding is characteristic.
    - Endoscopic ultrasound with FNA biopsy: this may detect metastatic disease and provide tissue sampling for histological diagnosis and molecular profiling. This is useful when neoadjuvant therapy is considered.
    - CT scan: an enhancing mass is characteristic; exophytic masses are not uncommon. Metastatic disease may be detected.
- The mainstay therapy includes surgical resection and treatment with tyrosine kinase inhibitors (TKIs). In low-risk tumors (small gastric GISTs with a low mitotic index), resection alone may be sufficient. In higher-risk cases, adjuvant, or neoadjuvant therapy with TKI is undertaken.
- TKIs bind to the tyrosine kinase receptor of GIST cells. They lead to inhibition of signaling pathways (MAPK, JAK/STAT, and PI3K/AKT), thus inhibiting cellular proliferation. The initial Imatinib was the initial TKI used for the treatment of GISTs. Other TKIs include sunitinib and regorafenib.
- Surgery is indicated for tumors > 2 cm. Tumor resection with clear gross margins is the main surgical objective. Usually lymphadenectomy is not required. Wedge resection of small gastric tumors is possible if anatomically and functionally feasible.
- A Whipple procedure (pancreaticoduodenectomy) may be indicated for duodenal tumor resection.
- Tumors < 2 cm may be removed endoscopically.
- GISTs are soft and friable; t is important to be careful during surgical manipulation of larger tumors. Fracture may lead to tumor spillage.
- Adjuvant treatment with TKIs is considered in tumors with a high-risk profile. Relapse rates and overall survival are lower with adjuvant imatinib when compared with surgery alone.
- Neoadjuvant TKI treatment is considered for locally advanced, borderline resectable, and unresectable lesions. It is also considered in tumors in precarious anatomic locations (gastroesophageal junction, ampullary region, and rectum). Tumor downsizing with TKI therapy may increase resectability and reduce operative morbidity. This is especially true for large rectal GISTs with KIT exon 11 mutations.

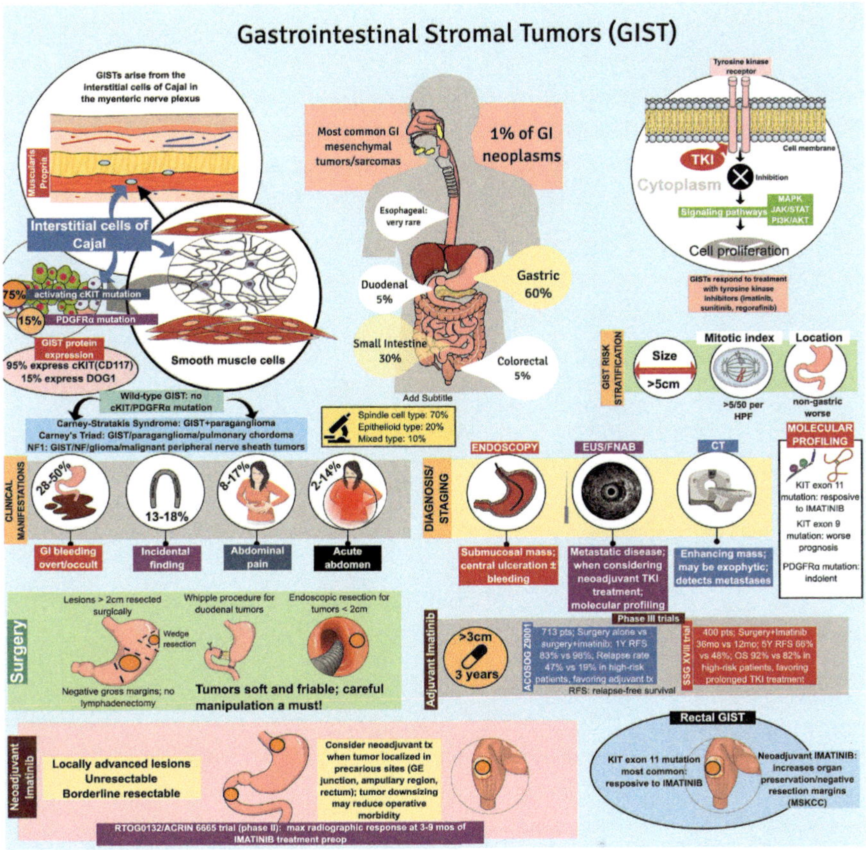

Fig. 1 Gastrointestinal stromal tumors.

Further reading

[1] Helmink BA, Bailey CE, Tarpley JL. The management of small bowel tumors. In: Cameron JL, Cameron AM, editors. Current surgical therapy. 12th ed. Philadelphia, PA: Elsevier; 2017. p. 121–7.

[2] Stettler G, McCarty MD. Gastrointestinal stromal tumors. In: McIntyre Jr RC, Schulick RD, editors. Surgical decision making. 6th ed. Philadelphia, PA: Elsevier; 2020. p. 256–8.

[3] Miettinen M, Sarlomo-Rikala M, Lasota J. Gastrointestinal stromal tumors: recent advances in understanding of their biology. Hum Pathol 1999;30:1213.

[4] Everett M, Gutman H. Surgical management of gastrointestinal stromal tumors: analysis of outcome with respect to surgical margins and technique. J Surg Oncol 2008;98:588.

CHAPTER 5

Gastrointestinal neuroendocrine (carcinoid) tumors

- Carcinoid tumors are currently termed **primary neuroendocrine tumors (NETs)**. NETs arise from enterochromaffin cells (Kulchitsky cells) of the aerodigestive tract. These tumors have the unique ability to synthesize and secrete bioactive substances. They are characterized by the expression of somatostatin receptors on their cell surface which has both diagnostic and therapeutic implications.
- The incidence of NETs is rising; currently they affect 4.7 people per 100,000. Median age of diagnosis is 60–70 years of age; African Americans are more frequently affected.
- NETs are divided into foregut NETs (bronchial, pulmonary, stomach duodenum, and proximal small bowel), midgut NETs (mid-jejunum to mid-transverse colon; appendix) and hindgut NETs (distal transverse colon, rectum). Pancreatic NETs were previously termed islet cell tumors.
- The distribution and frequency of NETs are as follows:
    - Broncopulmonary: 30%
    - Gastric: 7%
    - Small bowel (45%); these are the most common malignant small bowel tumors
    - Appendix: 16%
    - Colon: 11%
    - Rectum: 20%
- Gastric-duodenal NETs are divided into three types:
    - Type I (70%–80%): associated with chronic atrophic gastritis and pernicious anemia. These are characteristically multiple small tumors with central ulceration.
    - Type II (5%): these are gastrinomas and are associated with the Zollinger-Ellison syndrome, which will be discussed separately.
    - Type III (20%): These are aggressive tumors and metastasize frequently.
- Gastric NETs usually secrete histamine, causing atypical carcinoid syndrome.

Digestive System Malignancies
https://doi.org/10.1016/B978-0-323-98369-3.00019-8

- Midgut NETs frequently invade the mesentery, metastasize to lymph nodes and the liver, and present with carcinoid syndrome. The majority are located within 60 cm of the ileocecal valve. Clinical presentation includes pain, intussusception, and bowel obstruction.
- Appendiceal NETs are usually found incidentally in appendectomy specimens. They are mostly located in the distal third of the vermiform appendix.
- Hindgut NETs are nonsecretory, even when metastases have occurred. The cecum is the most common location of colorectal NETs. They are usually large (> 5 cm); metastases are found in 65% of patients. Large lesions may present with change in bowel habits, abdominal pain, anorexia, weight loss, and diarrhea.
- Rectal NETs are frequently incidental endoscopic findings in older patients and are small.
- The WHO pathological classification of NETs considers histologic grade (proliferative tumor activity/mitotic rate and Ki67 index), and differentiation.
- Carcinoid syndrome is a constellation of symptoms (flushing, diarrhea, bronchospasm, and chronic sequelae) mediated by various humoral factors elaborated by well-differentiated NETs of the GI tract and lungs, which synthesize, store, and release a variety of polypeptides, biogenic amines, and prostaglandins. It is associated with disseminated or metastatic disease.
- One of the key molecules in carcinoid syndrome is tryptophan, converted to 5-hydroxytryptophan (5-HT) by tryptophan hydroxylase. 5-HT is converted to serotonin by aromatic decarboxylase. Both enzymes are found in cells in most NETs.
- Serotonin stimulates intestinal secretion and inhibits absorption, causing diarrhea. This is a common symptom in carcinoid syndrome.
- Serotonin hypersecretion may lead to niacin deficiency and pellagra.
- Serotonin stimulates fibroblasts. This may lead to fibrosis of the valves in the right heart (tricuspid/pulmonary valves); this is termed carcinoid heart disease. In addition, mesenteric and retroperitoneal fibrosis may lead to mesenteric ischemia and hydronephrosis.
- Serotonin is metabolized into 5-HIAA by monoamine oxidase and is excreted in the urine. Assay of urine collected in 24-h may lead to confirmation of carcinoid syndrome. Serum chromogranin A/B/C is also used in the biochemical diagnosis of NETs.
- Foregut NETs lack aromatic decarboxylase and do not produce excess serotonin. Atypical carcinoid syndrome (flushing without diarrhea), induced by histamine, is characteristic of foregut NET.

- Flushing in carcinoid syndrome is episodic and frequently provoked my alcohol consumption, defecation, stress and palpation of the liver. It is induced by bradykinin, prostaglandins, tachykinin, kallikrein, histamine and substance P.
- Bronchospasm may occur during flushing episodes. This may be confused with asthma. Of note, beta-agonists should not be administered, they may trigger prolonged vasodilation and hypotension.
- Diagnostic modalities used for primary/metastatic GI NETs include CT/MRI scans of the chest/abdomen/pelvis. NETs are highly vascular lesions. CT/MRI enterography may also be used for diagnosis.
- Nuclear imaging is useful for diagnosis of metastatic diseases. Somatostatin receptor imaging (octreoscan) and 68-Ga DOTATATE PET/CT scans are the most common modalities used for localization.
- Type I, 2 gastric NETs may be treated with endoscopic resection if small. Type 3 lesions are usually treated with gastrectomy.
- Large ampullary NETs are treated with the Whipple procedure (pancreaticoduodenectomy).
- Appendiceal NETs < 1 cm in size, or lesions 1.1–1.9 cm and no mesoappendiceal invasion/positive or unclear margins are treated with appendectomy alone.
- Appendiceal lesions > 1 cm or lesions 1.1–1.9 cm with mesoappendiceal invasion/positive or unclear margins are treated with right hemicolectomy.
- Small bowel NETs are treated with segmental resection. A generous lymphadenectomy must be performed. The entire small bowel is explored, since multiple lesions are found in 25% of cases.
- Colonic NETs are treated with colectomy and a generous lymphadenectomy.
- Rectal NETs may be endoscopically resected when < 1 cm with favorable histology. Larger lesions with unfavorable histology are treated with radical rectal resection (low anterior resection or abdominoperineal resection).
- Surgical treatment of metastatic disease confined to the liver includes anatomic liver resection, nonanatomic wedge resection, and hepatic artery embolization.
- Carcinoid crises (flushing, diarrhea, tachyarrhythmia, bronchospasm, and mental status changes) are treated with octreotide either as a drip or subcutaneous injection. In patients with unresectable metastatic disease, octreotide or lantreotide may control symptoms and delay disease progression.
- In locally advanced or unresectable metastatic disease, treatment with the mTOR inhibitor everolimus is an option with favorable results.

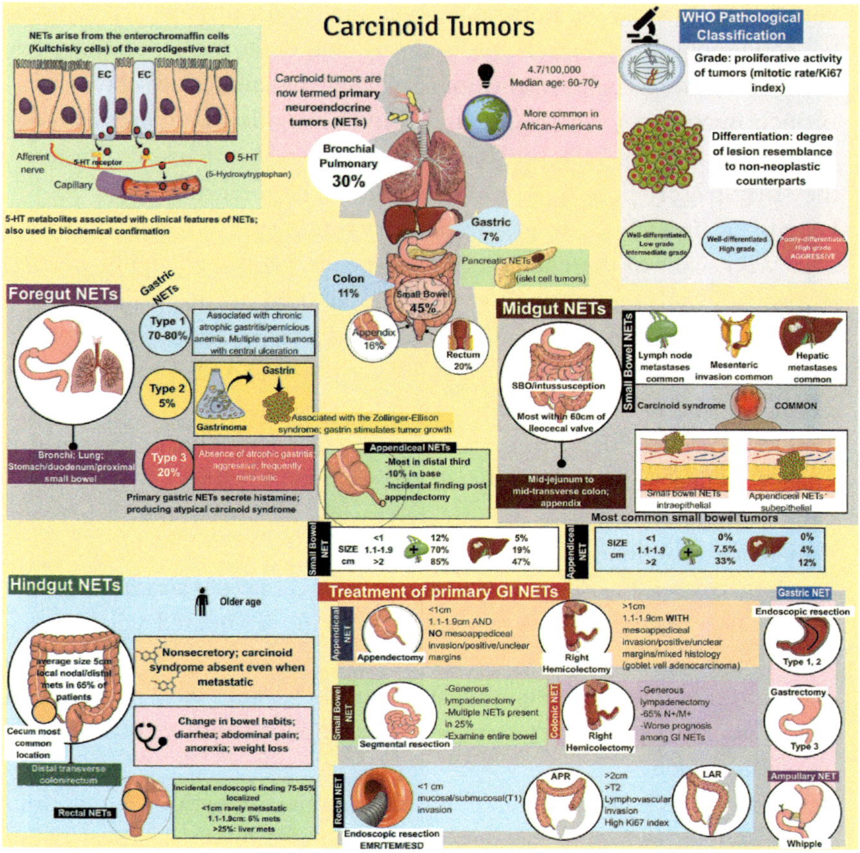

Fig. 1 Neuroendocrine tumors: Classification and treatment.

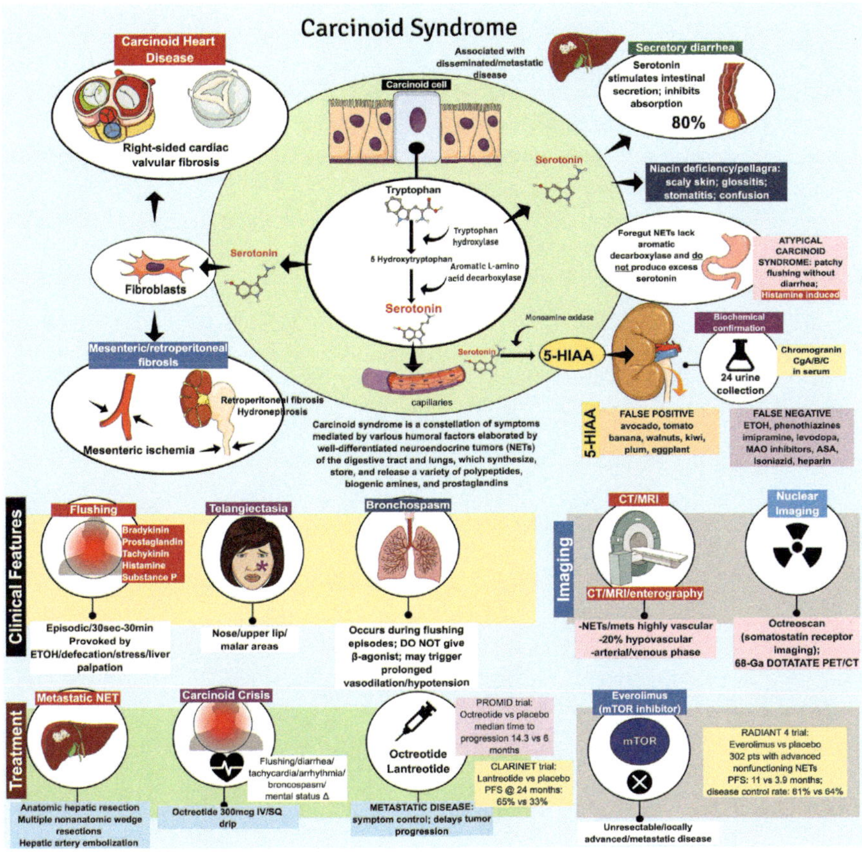

Fig. 2 Carcinoid syndrome.

Further reading

[1] Helmink BA, Bailey CE, Tarpley JL. The management of small bowel tumors. In: Cameron JL, Cameron AM, editors. Current surgical therapy. 12th ed. Philadelphia, PA: Elsevier; 2017. p. 121–7.

[2] O'Toole D, Grossman A, Gross D, et al. ENETS consensus guidelines for the standards of care in neuroendocrine tumors: biochemical markers. Neuroendocrinology 2009;90:194.

[3] Modlin IM, Lye KD, Kidd M. A 5-decade analysis of 13,715 carcinoid tumors. Cancer 2003;97:934.

[4] Jann H, Roll S, Couvelard A, et al. Neuroendocrine tumors of midgut and hindgut origin: tumor-node-metastasis classification determines clinical outcome. Cancer 2011;117:3332.

CHAPTER 6

Pancreatic neuroendocrine tumors

- Pancreatic neuroendocrine tumors (PNETs) were formerly termed as pancreatic islet tumors. These are rare tumors, comprising of 2%–3% of all pancreatic tumors. They derive from pluripotent pancreatic stem cells.
- PNETs can be classified as functional and nonfunctional. Functional tumors secrete hormones that result in a specific clinical presentation. Most (40%–75%) PNETs are nonfunctional.
- Hormones secreted by PNETs include glucose, insulin, glucagon, gastrin, vasoactive intestinal peptide, pancreatic polypeptide, and chromogranin A.
- Most PNETs are sporadic.
- PNETs can be associated with inherited disorders. They occur in multiple endocrine neoplasia type I (80%–100% of cases), Von-Hippel-Lindau disease (20% of cases), and tuberous sclerosis (1% of cases).
- The WHO pathological classification of PNETs takes into account tumor site, angioinvasion, and mitotic rate.
- In general, PNETs clinically present in three ways:
    - As a clinical syndrome associated with hormone secretion.
    - As a result of a mass effect caused by the tumor.
    - As an incidental finding on imaging.
- The most common diagnostic imaging modalities used are CT/MRI, endoscopic ultrasound (EUS), and nuclear imaging (octreoscan, 68-Ga DOTATATE PET/CT).
- Insulinomas are the most common functional PNETs (30%–45%). Most are benign; 10% are malignant. They are located within the head, body, or tail of the pancreas with equal distribution.
- Whipple's triad is characterized by neuroglycopenic symptoms (palpitations, sweating, tremors, headache, confusion, and vision disturbances), hypoglycemia (plasma glucose <40 mg/dL) and symptom relief with glucose administration.

Digestive System Malignancies
https://doi.org/10.1016/B978-0-323-98369-3.00017-4

- Biochemical confirmation is done with a 72-h fast during which glucose, c-peptide, proinsulin, and insulin levels are drawn every 6 hours until symptoms appear. Absence of elevated c-peptide/proinsulin levels suggests surreptitious antidiabetic drug abuse.
- Imaging for insulinoma includes MRI/CT scan (10%–40% sensitivity), octreoscan (25% sensitivity), EUS (70%–95% sensitivity), and selective angiography after stimulation with calcium infusion. Most lesions are <2 cm.
- Surgical treatment of insulinoma includes lesion enucleation for small tumors and resection for larger tumors. This can be performed with open or minimally invasive techniques. Lesions not found on preoperative imaging can be detected intraoperatively with intraoperative ultrasound or manual palpation.
- Metastatic insulinoma is rare. Treatment is symptomatic may include administration of diazoxide, octreotide, lantreotide, and glucocorticoids.
- Glucagonomas are malignant in 50%–80% of cases. They have equal distribution within the pancreas. Serum glucagon is elevated. Clinically, they present with the "4Ds"o: dermatitis, diabetes, diarrhea, and deep venous thrombosis. Necrolytic migratory erythema of the legs, trunk and/or perineum is characteristic. Surgical resection is the treatment of choice; octreotide or lantreotide can be used for symptomatic palliation.
- VIPoma, or tumors secreting vasoactive intestinal peptide, present with Werner-Morrison's syndrome (watery diarrhea, hypokalemia, and achlohydria). Most tumors are malignant and are distributed in the pancreatic body and tail. Serum VIP is usually >200 pg/mL. Metastases are common. VIPomas express somatostatin receptors in 90%; octreoscan is sensitive for tumor detection.
- Patients with VIPoma are initially treated with correction of their volume and electrolyte status. Resection and/or surgical debulking may improve symptoms; complete resection is ideal, but not always possible. Chemotherapy and/or chemoembolization may also offer symptomatic treatment.
- Nonfunctional PNETs do not produce clinical syndromes. They may present with abdominal pain, anorexia, weight loss or jaundice as a result of a mass effect on adjacent organs. Incidental finding on imaging is another mode of presentation. Despite the absence of a clinical syndrome, chromogranin A and pancreatic polypeptide are elevated in

these tumors and are used to assess response to therapy. Octreoscan is sensitive for in detection of these lesions. Surgical resection or debulking is performed when possible. Chemotherapy and/or chemoembolization may also offer symptomatic treatment.

- Metastatic disease (mostly hepatic) can be treated with resection, hepatic arterial embolization, chemoembolization, radiofrequency embolization, yttrium90 microsphere radioembolization, peptide receptor radionuclide therapy, and liver transplantation in highly selected cases.
- Cytoreductive surgery involves resection of the primary tumor and metastases when possible. Increased survival is achievable when 90% of tumor burden is resected when compared with nonsurgical therapy.
- Systemic treatments include lantreotide, streptozocin + 5FU/doxorubicin, and everolimus (an mTOR receptor inhibitor).

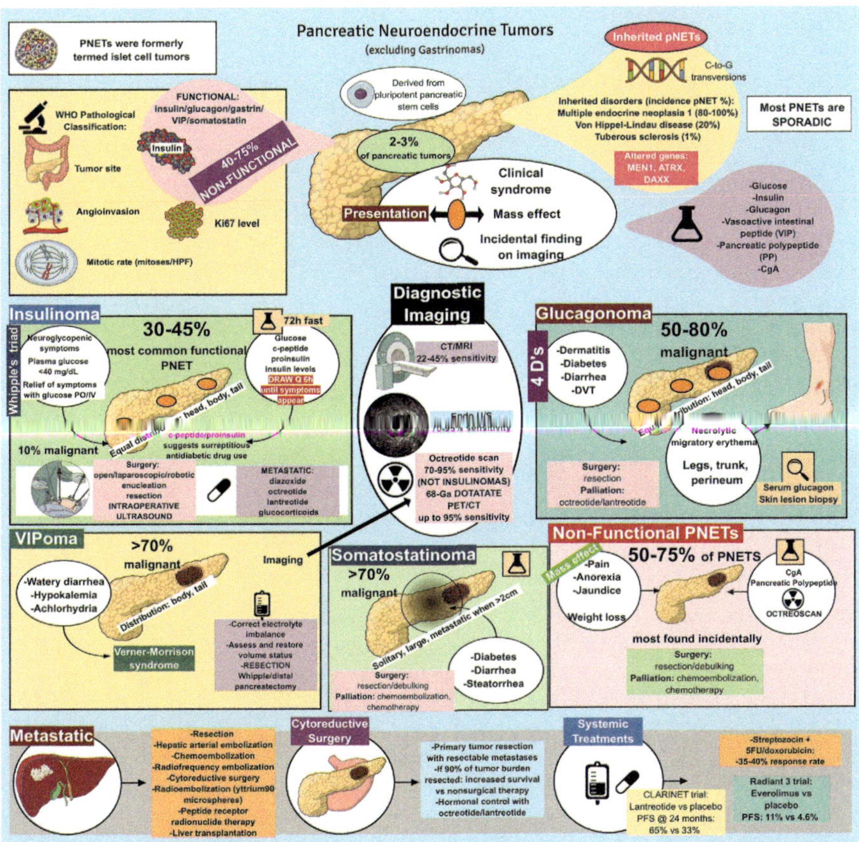

Fig. 1 Pancreatic neuroendocrine tumors.

Further reading

[1] Lou I, Chen H. The management of pancreatic islet tumors excluding gastrinomas. In: Cameron JL, Cameron AM, editors. Current surgical therapy. 12th ed. Philadelphia, PA: Elsevier; 2017. p. 121–7.

[2] Torphy RJ, Edil BH. Pancreatic neuroendocrine tumors tumors. In: McIntyre Jr RC, Schulick RD, editors. Surgical decision making. 6th ed. Philadelphia, PA: Elsevier; 2020. p. 260–2.

CHAPTER 7

Gastrinoma (Zollinger-Ellison syndrome)

- The Zollinger-Ellison syndrome (ZES) is characterized by gastric acid hypersecretion leading to severe acid-related peptic ulcer disease with diarrhea, caused by a gastrinoma, which is a neuroendocrine tumor.
- Gastrinomas occur in patients 20–50 years of age. The incidence is 0.5–2/100,000 and they are more common in males.
- The majority of gastrinomas are sporadic (80%). Sporadic gastrinomas metastasize more frequently.
- Gastrinomas may be associated with MEN1 (hyperparathyroidism, pituitary tumors, and gastrinoma) in 20% of cases. MEN1 gastrinomas have a more favorable prognosis.
- The gastrinoma triangle (sweep of duodenum/head of pancreas) is the location of 80% of these tumors. The majority of gastrinomas are located in the duodenum.
- Duodenal gastrinomas are usually < 1 cm and metastasize less frequently than pancreatic gastrinomas.
- Ectopic gastrinomas are found in the stomach, peripancreatic lymph nodes, liver and bile ducts, ovaries, and in the heart.
- The pathophysiology of ZES is the result of gastrin hypersecretion from tumor cells. Hypergastrinemia exerts a trophic effect on gastric parietal cells, leading to excess acid secretion. Acid hyposecretion causes peptic ulceration. In addition, the excess acid is not reabsorbed in the small bowel and colon; bicarbonate neutralization is insufficient. The result is sodium/water nonabsorption of the gut and chronic diarrhea.
- Clinical manifestations of ZES include peptic ulcer disease refractory to treatment with diarrhea.
- Diarrhea improved with PPI treatment may hint at the diagnosis of ZES.
- Delay in diagnosis of ZES is common; mean time to diagnosis may be 5–8 years.

Digestive System Malignancies
https://doi.org/10.1016/B978-0-323-98369-3.00013-7

- Most ulcers in ZES are solitary, located in the first portion of the duodenum. Ulcers in unusual locations (distal duodenum, proximal jejunum), and multiple ulcers refractory to treatment, are highly suspicious for ZES.
- A fasting serum gastrin level (100–1000 pg/mL) in the setting of low gastric pH (<2) is diagnostic.
- The secretin stimulation test may be used to confirm the diagnosis. After cessation of antiulcer medications prior to the test, IV secretin is administered, and serum gastrin is measured at intervals (2, 5, and 10 min). Elevation of gastrin above baseline confirms the diagnosis.
- Gastric acid secretion studies (basal acid output >15 mEq/h) are no longer used.
- Secondary hypergastrinemia may occur in the following:
  - Retained antrum (after gastrectomy)
  - Antral G cell hyperplasia (the secretin stimulation test does not lead to hypergastrinemia)
  - *Helicobacter pylori* pangastritis
  - Renal failure
  - PPIs
- Upper endoscopy reveals peptic ulceration and prominent gastric folds.
- CT/MRI is used to image larger lesions and metastatic disease.
- EUS in useful for detecting pancreatic gastrinomas. In addition, FNA is possible.
- Nuclear imaging is sensitive in detecting gastrinomas.
- Other modalities include selective angiography with arterial gastrin stimulation, followed by venous sampling of gastrin. This is laborious and rarely used. Operative exploration with palpation, duodenotomy, and direct mucosal inspection is used when imaging does not reveal the tumor.
- Surgical resection is the primary treatment of choice for sporadic gastrinoma. Enucleation of small pancreatic tumors, duodenotomy, and enucleation of small duodenal tumors are the most common procedures performed. Peripancreatic and peri-duodenal lymph node sampling is recommended; lymphadenectomy leads to better biochemical cure.
- Larger lesions may require distal pancreatectomy or pancreaticoduodenectomy (Whipple procedure).
- A significant proportion of resected patients will require additional antisecretory therapy.
- Primary treatment of MEN1 gastrinomas is medical; omeprazole and pantoprazole are commonly used. Octreotide/lantreotide are given if symptoms persist.

- Larger (> 2 cm) MEN1 gastrinomas are surgically resected. Subtotal parathyroidectomy should precede the resection of gastrinoma.
- Gastrinomas metastasize to the liver, spine, and pelvis. Somatostatin receptor analogs are used for symptom control. These do not have antitumor activity.
- Liver resection, hepatic artery embolization and/or chemoembolization, microsphere embolization, radiofrequency or cryoablation, and liver transplantation all have been utilized for metastatic disease.

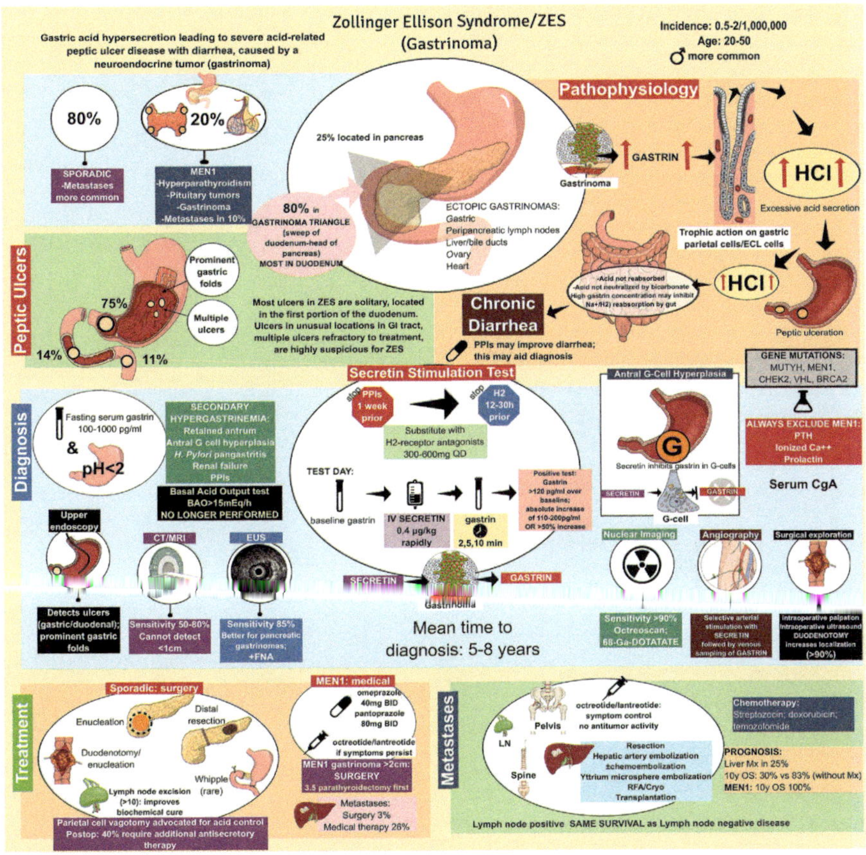

Fig. 1 Gastrinoma.

Further reading

[1] Norton JA, Krampitz GW, Jensen RT. The management of the Zollinger-Ellison syndrome. In: Cameron JL, Cameron AM, editors. Current surgical therapy. 12th ed. Philadelphia, PA: Elsevier; 2017. p. 86–91.

[2] El Lakis M, Kebebew E. Zollinger-Ellison syndrome. In: McIntyre Jr RC, Schulick RD, editors. Surgical decision making. 6th ed. Philadelphia, PA: Elsevier; 2020. p. 348–51.

CHAPTER 8

Pancreatic adenocarcinoma

- Pancreatic ductal adenocarcinoma (PDAC) is the fourth overall cause of cancer-related deaths, and the second cause of digestive tract-related deaths. Approximately 15%–20% of patients have potentially resectable tumors. Overall 5-year survival is 10%–30%. Ninety-five percent of pancreatic neoplasma are adenocarcinomas.
- Risk factors include smoking, older age, and male sex.
- Familial risk factors associated with pancreatic adenocarcinoma include:
  - BRCA 1,2
  - FAMMM (Familial atypical multiple mole melanoma) syndrome
  - Familial adenomatous polyposis
  - Peutz-Jegher's syndrome
  - Ataxia-telangiectasia
  - Li-Fraumeni syndrome
  - Hereditary pancreatitis
  - Familial pancreatic cancer (in which two first-degree relatives are affected)
- The majority (60%–70%) of pancreatic adenocarcinomas are located in the pancreatic head/uncinate process. They originate from the pancreatic ductal and acinar cells.
- Periampullary neoplasms include PDAC, distal common bile duct cancer, carcinoma of the ampulla of Vater, and duodenal adenocarcinoma. Clinical presentation is similar.
- Clinical presentation includes abdominal pain, obstructive jaundice, and weight loss. Abdominal pain is insidious, visceral, and may radiate to back and flank.
- Patients with pancreatic head lesions present with jaundice, those with body/tail lesions present mostly with central abdominal pain radiating to back without jaundice.
- New-onset diabetes may precede PDAC in 25% of patients.
- Courvoisier's sign (jaundice with palpable gallbladder) is found in 13% of patients.

Digestive System Malignancies
https://doi.org/10.1016/B978-0-323-98369-3.00021-6

- Palpable periumbilical, supraclavicular nodes, and ascites are signs of advanced disease.
- Schmidt's triad is characteristic of acinar cell pancreatic adenocarcinoma and includes panniculitis, polyarthralgia and eosinophilia.
- Treatment planning for PDAC is centered on determining tumor resectability. In most centers with high volumes and experience in treating PDAC, histologic confirmation is not mandatory. Several diagnostic modalities are available for pretreatment assessment.
- Transabdominal ultrasound detects biliary tract dilation and is sensitive for large pancreatic masses.
- Triple-phase contrast-enhanced multi-detector CT scan using a "pancreatic protocol" may depict a hypoattenuating pancreatic mass, the "double-duct" sign, and hepatic metastases.
- MRCP may also depict pancreatic masses, biliary tree anatomy, and metastatic disease. It is not used as frequently as CT.
- ERCP detects suspicious pancreatic duct stricture (> 1 cm), and superimposing strictures. Brush/forceps cytology is possible (sensitivity 50%–60%). In addition, a stent may be deployed for preoperative relief of obstructive jaundice or palliation.
- Biopsy is indicated in the following circumstances:
    - Evidence of systemic spread-plan for chemotherapy without surgery
    - Patients unfit for surgery
    - Planned neoadjuvant therapy
    - When alternative diagnoses are considered (e.g., focal chronic pancreatitis, autoimmune pancreatitis)
- The tumor marker CA 19-9 is not diagnostic and is used in follow-up or for assessing response to adjuvant/neo-adjuvant therapy.
- Patients fit for surgery, with a potentially resectable lesion suspicious for pancreatic cancer can proceed to operation without biopsy.
- CT-scan is highly predictive of **unresectable** disease. It is not predictive for **resectable** disease. Approximately 30% of patients deemed resectable on imaging are subsequently found to be unresectable at operation.
- Staging laparoscopy is used for detection of unresectable disease (liver metastases, peritoneal surface malignancy, and positive peritoneal washing cytology) at several centers. This is performed selectively in patients with tumors > 3 cm, body/tail tumors and in patients with CA-19-9 levels > 100 U/mL

- Borderline resectable tumors are characterized by:
  - Absence of distant metastases.
  - Short-segment thrombotic occlusion of the portal vein/superior mesenteric vein which is reconstructable with vascular anastomosis/ interposition.
  - Short-segment tumor encasement of the portal vein/superior mesenteric vein which is reconstructable with vascular anastomosis/ interposition.
  - Tumor involving <180 degrees of portal vein/superior mesenteric vein.
  - Tumor abutting <180 degrees of superior mesenteric artery circumference.
- Pancreaticoduodenectomy (Whipple procedure) is the mainstay of surgical treatment for pancreatic head/uncinate lesions. The head of pancreas, duodenum, and proximal jejunum are resected. Reconstruction includes pancreatojejunostomy, hepaticojejunostomy, and gastrojejunostomy. Pylorus preserving Whipple procedures have a lower incidence of "dumping," marginal ulceration, and bile reflux gastritis than standard procedures (in which an antrectomy is performed).
- Whipple procedures have relatively high morbidity. Major complications include delayed gastric emptying, pancreatojejunostomy leak, and delayed postpancreatectomy hemorrhage. Operative mortality is approximately 3% at high-volume centers. Five-year overall survival after Whipple is 10%–25%. Median survival is approximately 10–20 months.
- In patients with highly elevated bilirubin and potential delay in surgical treatment, preoperative biliary drainage is considered.
- Distal pancreatectomy with or without splenectomy is performed in patients with resectable lesions of the pancreatic body/tail.
- Most patients operated for PDAC upfront will be advised to undergo adjuvant chemotherapy or chemoradiotherapy. Several clinical trials have demonstrated improved survival with adjuvant chemotherapy (mostly with gemcitabine). In most European and UK centers, chemotherapy is used; in North America, adjuvant chemoradiotherapy is utilized.
- Neoadjuvant chemotherapy/chemoradiotherapy is considered in patients with borderline-resectable lesions. Using this approach, all patients receive systemic therapy, R1 resections are less frequent, pancreatic fistulae have been less frequent, and tumor downsizing is possible. In addition, patients with aggressive disease are identified, thus can avoid futile surgery.

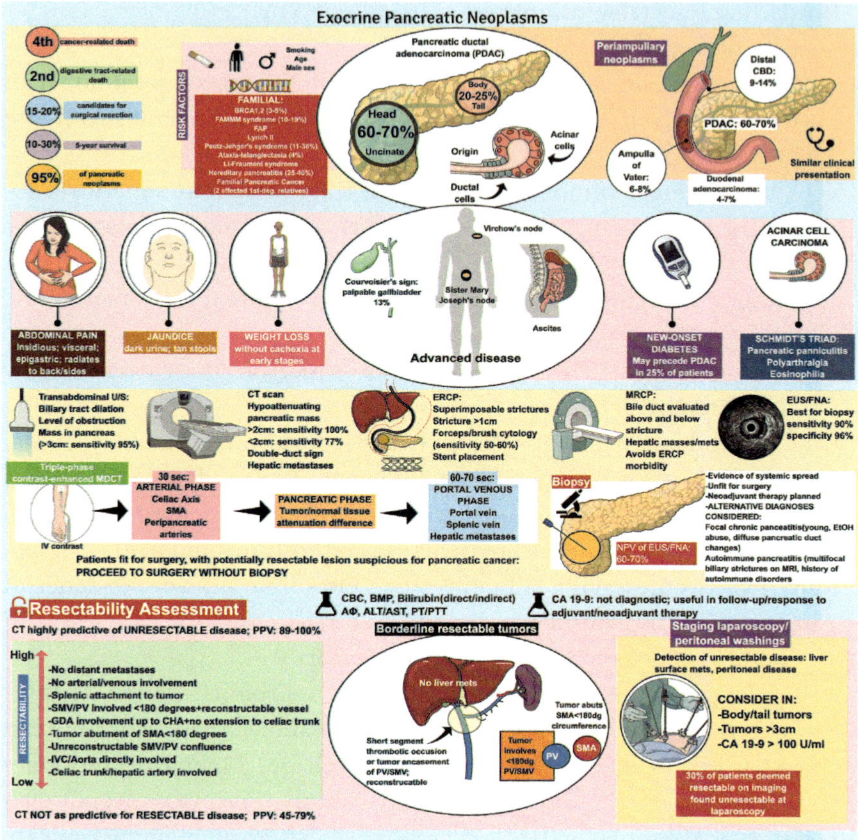

Fig. 1 Exocrine pancreatic neoplasms: Epidemiology and evaluation.

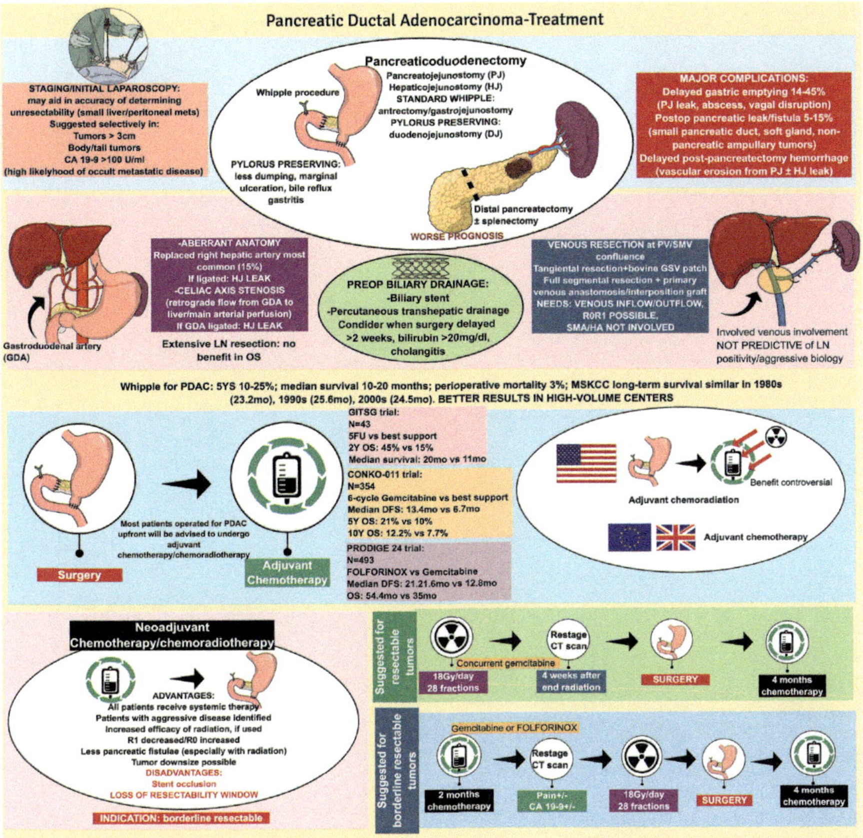

Fig. 2 Exocrine pancreatic neoplasms: Treatment.

Further reading

[1] Griffin JF, Poruk KE, Wolfgang CL. In: Cameron JL, Cameron AM, editors. Current surgical therapy. 12th ed. Philadelphia, PA: Elsevier; 2017. p. 538–48.

[2] Fong ZV, Lillemoe KD. Periampullary carcinoma. In: McIntyre Jr RC, Schulick RD, editors. Surgical decision making. 6th ed. Philadelphia, PA: Elsevier; 2020. p. 264.

[3] He J, Cameron JL. Pancreatic mass. In: McIntyre Jr RC, Schulick RD, editors. Surgical decision making. 6th ed. Philadelphia, PA: Elsevier; 2020. p. 268.

[4] Khorana AA, Mangu PB, Berlin J, et al. Potentially curable pancreatic cancer: American Society of Clinical Oncology Clinical Practice Guideline. J Clin Oncol 2016;34:2541.

[5] Ryan DP, Hong TS, Bardeesy N. Pancreatic adenocarcinoma. N Engl J Med 2014;371:1039.

CHAPTER 9

Pancreatic cystic lesions

- Pancreatic cystic lesions are rare, comprising 1%–2% of pancreatic neoplasms, and 50% of pancreatic cystic lesions.
- The main types of pancreatic cystic lesions are:
  - Serous cystic neoplasms (16%)
  - Mucinous cystic lesions (23%)
  - Solid pseudopapillary neoplasms (3%)
  - Intraductal papillary mucinous neoplasms-IPMN (38%)
- Serous cystic neoplasms are well-demarcated multicystic lesions. On CT a central scar or sunburst lesion is pathognomonic. The microcystic variant may mimic solid lesions on lesions on imaging. Pancreatic serous cystic lesions are nearly never malignant.
- Serous cystic neoplasms are usually asymptomatic; large lesions may cause mass effect symptoms, including abdominal discomfort/pain, gastric outlet, or bile duct obstruction.
- Mucinous cystic lesions occur more frequently in females. They present as unilocular or as separated lesions on imaging. Their pathology is characterized by an ovarian-type stroma or cyst lined by mucinous epithelium. Mucinous cystic lesions are premalignant. A thick, irregular, or calcified cystic wall with an internal solid component is highly suspicious for malignancy.
- Solid pseudopapillary lesions are mixed cystic/solid lesions with the "4Cs": circumscribed, cystic appearing, with a capsule, and calcification.
- IPMN can be classified as main-duct IPMN (MD-IPMN), branch-duct (BD-IPMN), and mixed type (combination of MD/BD).
- In MD-IPMN, the main pancreatic duct may be diffusely or partially dilated, especially in the pancreatic head/uncinate process. Other findings include abrupt changes in duct diameter with distal pancreatic atrophy, and solid intraductal nodules. A cystic lesion in the pancreatic head is an IPMN until proven otherwise. MD-IPMN are malignant in 70% of cases

Digestive System Malignancies
https://doi.org/10.1016/B978-0-323-98369-3.00007-1

- In BD-IPMN, branch duct dilatation is observed. This may appear as multiple small cysts off the main pancreatic duct. They are malignant in 20% of cases.
- There are four histological subtypes of IPMN: gastric, intestinal, pancreatobiliary, and oncocytic.
- Extrapancreatic cancers may occur in 10%–45% of patients with IPMN. These include breast, colon, and prostate cancer.
- Cross-sectional imaging (MRI/MRCP, and triple-phase MDCT) is instrumental in the diagnosis and classification of pancreatic cystic lesions.
- The Sendai criteria are used for risk stratification of IPMNs to assist in treatment decision-making. The criteria are based on high-quality cross-sectional imaging findings.
- High-risk Sendai criteria include:
  - Main pancreatic duct > 10 mm
  - Presence of an enhancing solid component
  - Obstructive jaundice
- When high-risk criteria are present, patients proceed to surgery without further diagnostic evaluation.
- "Worrisome" Sendai features include:
  - Tumor > 3 cm
  - Main duct 5–9 mm
  - Presence of a nonenhancing mural nodule
  - Abrupt change in ductal diameter with distal pancreatic atrophy
  - Thick or enhancing cyst walls
- When "worrisome" features are present, EUS/FNA is recommended. This provides sampling for cytology, tumor markers, and molecular markers. Information from EUS/FNA may determine the type of pancreatic cystic lesion type, which may facilitate decision-making.
- ERCP may also provide material for cytology and delineate the pancreatic duct anatomy. An enlarged, gaping ampulla (fish mouth or fish-eye sign) is pathognomonic for the diagnosis of IPMN.
- Surgical resection of serous cystic neoplasms is undertaken for symptomatic lesions.
- Mucinous cystic lesions and solid pseudopapillary lesions are resected.
- Resection is indicated for IPMNs with high risk or "worrisome" Sendai criteria.
- Pancreatic head lesions are treated with pancreaticoduodenectomy. Distal pancreatectomy is performed for lesions in the body/tail. Total pancreatectomy may be indicated for mixed IPMNs.

- Surveillance of IPMNs may be recommended in patients with inconclusive findings on EUS/FNA, or small lesions (< 2 cm).

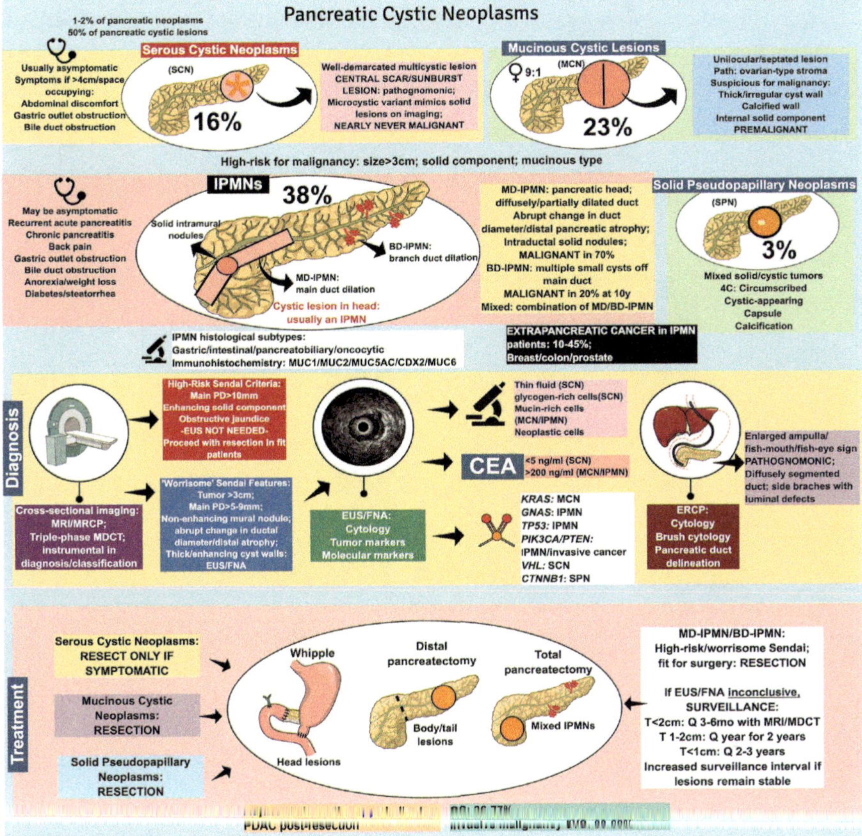

Fig. 1 Pancreatic cystic neoplasms: Classification, diagnosis, and treatment.

Further reading

[1] Datta J, Vollmer Jr CM. Unusual pancreatic tumors. In: Cameron JL, Cameron AM, editors. Current surgical therapy. 12th ed. Philadelphia, PA: Elsevier; 2017. p. 565–71.

[2] Dimou FM, Perone JA, Riall TA. Intraductal papillary mucinous neoplasms of the pancreas. In: Cameron JL, Cameron AM, editors. Current surgical therapy. 12th ed. Philadelphia, PA: Elsevier; 2017. p. 572–9.

[3] Allen PJ. Pancreatic cystic neoplasms. In: McIntyre Jr RC, Schulick RD, editors. Surgical decision making. 6th ed. Philadelphia, PA: Elsevier; 2020. p. 266.

CHAPTER 10

Hepatocellular carcinoma

- Hepatocellular carcinoma (HCC) constitutes for 80% of primary hepatic tumors. Worldwide, 700,000 cases are reported annually. An increase in the incidence has been observed. HCC is more frequent in females.
- Most cases of HCC develop in cirrhotic patients. In Western countries, these tumors develop in cirrhosis secondary to alcohol abuse and hepatitis C; in Asia, cirrhosis secondary to hepatitis B and C is the most common cause. The annual incidence of HCC in cirrhotics is 3%–5%.
- Disorders associated with increased risk for HCC include:
  - Tyrosinemia
  - α1-antitrypsin deficiency
  - Wilson's disease
  - Hypercitrullinemia
  - Porphyria cutanea tarda
  - Hemochromatosis
- Exposure to several chemicals/drugs/toxins may increase risk for HCC:
  - Thorotrast
  - Vinyl chloride monomers
  - Arsenic
  - Aflatoxins
  - Chronic anabolic steroid use
- Symptoms in patients with HCC include jaundice, weight loss, ascites, and the sequelae of portal hypertension (esophageal varices, splenomegaly/hypersplenism).
- Paraneoplastic syndromes are frequent in patients with HCC. These may include erythrocytosis, bone lesions (hypercalcemia, osteolysis), and skin lesions (pemphigus, pityriasis rotunda, dermatomyositis).
- Alpha-fetoprotein and des-gamma-carboxyprothrombin are elevated in HCC. A full metabolic panel, and complete blood count are also obtained. Initial imaging is performed with transabdominal ultrasound. A positive ultrasound finding with elevated α-fetoprotein increases diagnostic sensitivity for HCC.
- Four-phase (nonenhanced, arterial, portal venous, and delayed phase) MDCT is performed if ultrasound findings are positive. HCC is characteristically hyperattenuated in the arterial phase with rapid washout.

Digestive System Malignancies
https://doi.org/10.1016/B978-0-323-98369-3.00004-6

- MRI depicts a hyperintense lesion on T1-weighted images.
- Regarding PET/CT, 50% of HCC is not sensitive to FDG and may not be depicted.
- Treatment planning is based on staging, classification, and the Child-Pugh/MELD score.
- The Barcelona Clinic Liver Cancer System classifies HCC patients by taking into account of the lesion, hepatic function, and patient performance status. Patients with HCC are staged as having early, intermediate, advanced, and end-stage disease. Treatment consists of resection/transplantation, transarterial embolization/chemoembolization, and sorafenib and palliative care, respectively.
- Li-RADS is an imaging classification system that use findings on CT/MRI.
- Ideal candidates for HCC resection are the patients with the following:
  - Child's-Pugh A
  - Early Barcelona score
  - Single lesions
  - No metastases
  - Feasible R0 resection
  - Fibrolamellar type of HCC
- Principles of hepatic resection include negative margins, adequate amount of functional parenchyma after resection, intact hepatic arterial and portal venous inflow, intact hepatic venous outflow, and adequate biliary drainage. Anatomic resection includes sectionectomy, segmentectomy, and subsegmentectomy.
- Liver transplantation is a therapeutic option for HCC. Both HCC and cirrhosis are treated. The Milan criteria for transplantation in patients with HCC include:
  - A single tumor <5 cm
  - Up to 3 lesions <3 cm
  - Absence of gross vascular invasion
  - Absence of spread outside liver
- Liver transplantation may be associated with 1-year survival rates of 40%–70% and 5-year survival rates of 20%–45%.
- Radiofrequency ablation (RFA) of HCC can be performed via the percutaneous, laparoscopic or open routes. Coagulative necrosis of the tumor is undertaken. Indications for RFA include:
  - Child's-Pugh A/B patients
  - Tumors <4 cm
  - Cirrhotic patients which are poor operative candidates
  - Bridge to transplantation

- In small tumors (<3 cm), RFA outcomes are comparable to HCC resection. Incomplete ablation is frequent in larger tumors, with high recurrence rates as a result.
- Transarterial chemoembolization (TACE) and embolization (TAE) are other therapeutic modalities. Ischemic necrosis of HCC is induced by embolizing the hepatic artery tributaries feed in the tumor. PVC, gelatin, ivalon, and lipiodol are frequently used. Chemotherapeutic drugs such as adriamycin, cisplatin, and mitomycin-C are used in TACE.
- TACE/TAE are considered as the standard of care for Barcelona intermediate-stage HCC. It is also useful for tumor downsizing (in the face of possible resection), treatment of large tumors, multifocal disease, as well as a bridge to transplantation.
- Combined treatment with TACE and RFA has better outcomes than either treatment alone.
- In advanced HCC, ethanol ablation and sorafenib (an oral VEGF inhibitor) have been used for palliative therapy.

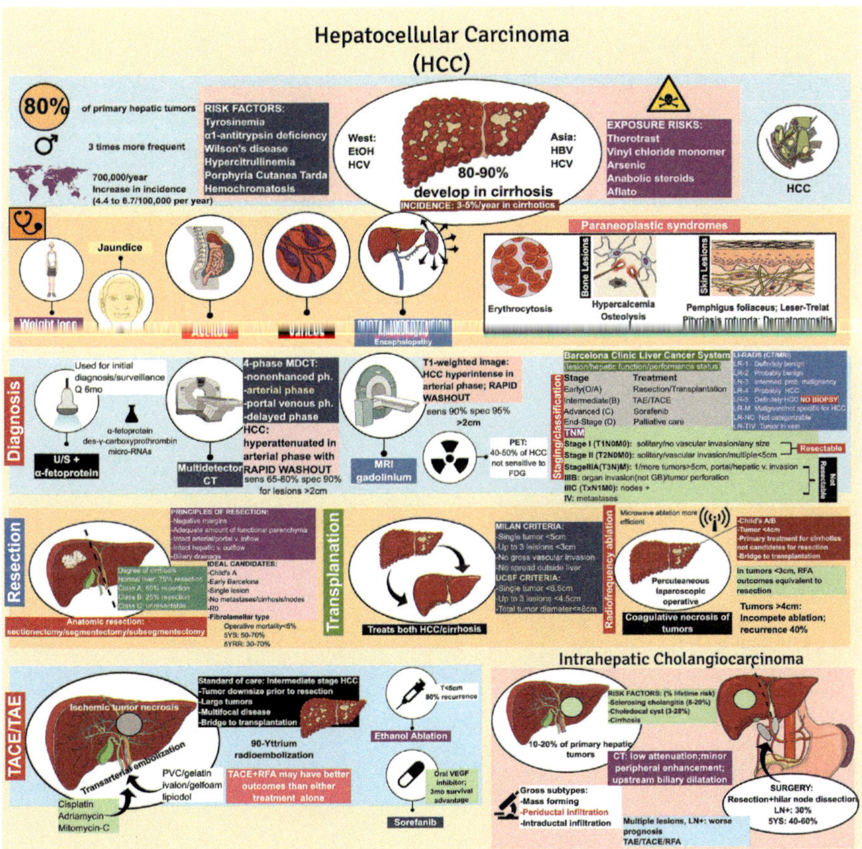

Fig. 1 Hepatocellular carcinoma.

Further reading

[1] Schulick RD, Gleisner AL. The management of malignant liver tumors. In: Cameron JL, Cameron AM, editors. Current surgical therapy. 12th ed. Philadelphia, PA: Elsevier; 2017. p. 360–5.

[2] Fong ZV, Tanabe KK. Ablation of colorectal carcinoma liver metastases. In: Cameron JL, Cameron AM, editors. Current surgical therapy. 12th ed. Philadelphia, PA: Elsevier; 2017. p. 371–5.

[3] Beard RE, Geller DA. Hepatocellular carcinoma. In: McIntyre Jr RC, Schulick RD, editors. Surgical decision making. 6th ed. Philadelphia, PA: Elsevier; 2020. p. 248–51.

CHAPTER 11

Metastatic colorectal carcinoma (CRC)

- Hepatic metastases of colorectal carcinoma (CRC) are the most common hepatic tumors. Approximately 30%–50% of patients diagnosed with CRC will have synchronous or metachronous hepatic metastases.
- An increase of carcinoembryonic antigen (CEA), AST/ALT, bilirubin, or alkaline phosphatase during posttreatment surveillance for CRC is suggestive of local or distant metastatic disease.
- CT scan (4-phase) is the primary imaging modality; hepatic metastases appear as hypodense lesions in the venous phase of the scan. MRI may distinguish tumors from cysts and is better for detection of metastases in steatotic livers.
- PET-CT is useful for detecting extrahepatic metastatic disease. Chemotherapy may decrease sensitivity and metastasis detection rates.
- The surgical strategy in patients with CRC and synchronous hepatic metastases is individualized for each patient. The appropriate sequence and timing of chemotherapy, hepatic resection, and colorectal resection are not clearly defined.
- Most frequently, upfront chemotherapy is given. This is useful for tumor response assessment and treatment of micrometastases.
- Concomitant colorectal and hepatic tumor resection may be performed if minor liver resection and a straightforward colorectal resection is feasible.
- "Liver first" resection is appropriate in patients with extensive hepatic metastatic disease, or in patients with rectal cancer with planned neoadjuvant chemoradiotherapy.
- A "colorectal first" resection is indicated for symptomatic CRC; namely, bleeding and/or obstruction.
- Prognostic factors associated with metastatic CRC include size of largest hepatic tumor, the disease-free interval between primary tumor

Digestive System Malignancies
https://doi.org/10.1016/B978-0-323-98369-3.00012-5

and metastatic disease (shorter interval = worse prognosis), the presence of extrahepatic metastases, and the lymph node status of the primary tumor.

- The future liver remnant (FLR) assessment is important in the preparation for hepatic resection. Liver volumetry is performed with CT scanning. In a healthy liver, a postresection FLR = 20% of total liver volume (TLV) is acceptable. When preoperative chemotherapy has been administered, an FLR = 30% of TLV is acceptable. Finally, in cirrhotic livers, an FLR = 40% of TLV is acceptable.
- If FLR is found to be insufficient, portal venous embolism of feeding branches to the tumor can be performed to induce contralateral-side hepatic growth. An interval of 4–6 weeks postembolization is recommended prior to resection.
- Hepatic resection, when feasible, provides optimal oncological outcomes. Parenchymal-sparing resections are preferred; they offer similar recurrence and overall survival rates when compared with major hepatectomy. Resection combined with ablative modalities is possible with multiple metastases.
- Radiofrequency ablation (RFA) of metastatic CRC can be performed via the percutaneous, laparoscopic or open routes. Microwave ablation is an option.
- Hepatic artery infusional chemotherapy using FUDR (floxuridine) via a hepatic artery infusion pump is a well-known treatment modality. The main tumor blood supply is from the hepatic artery.
- Additional/alternative treatments may include cytotoxic chemotherapy with FOLFOX/FOLFIRI, radioembolization with yttrium-90-microspheres, and the targeted therapy with monoclonal antibodies.
- Bevacizumab, an anti-VEGF antibody is widely used. Cetuximab, an anti-EGFR antibody is used in metastatic CRC with wild-type *KRAS*. Pembrolizumab reverses tumor-induced T-cell dysfunction; CRC with microsatellite instability is responsive to treatment with this agent.

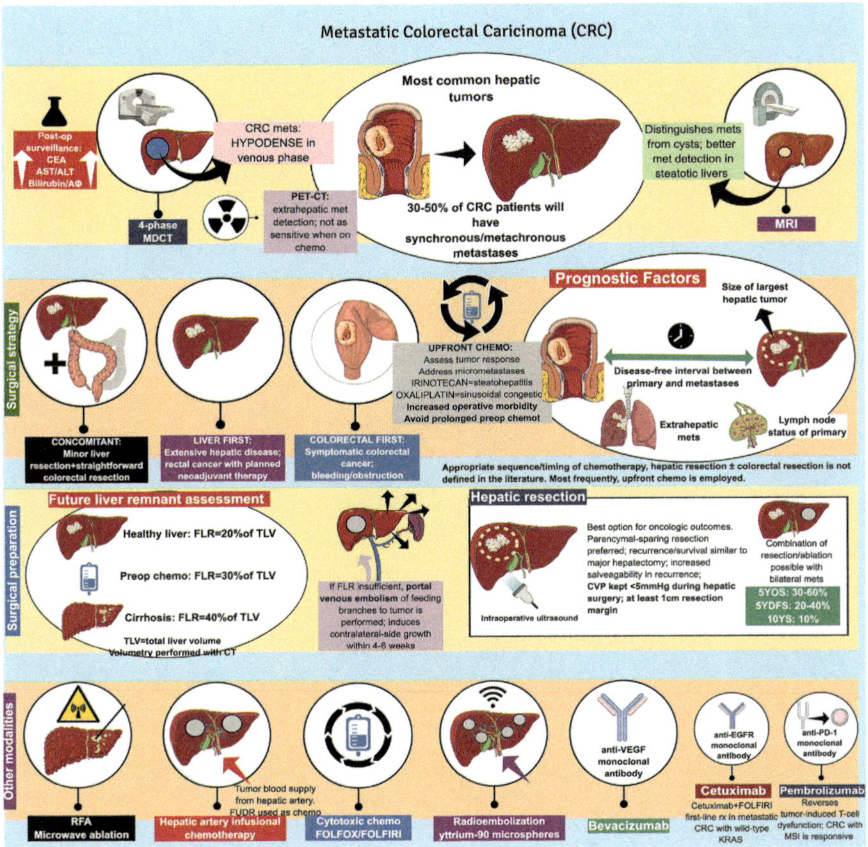

Fig. 1 Metastatic colorectal carcinoma.

Further reading

[1] Schulick RD, Gleisner AL. The management of malignant liver tumors. In: Cameron JL, Cameron AM, editors. Current surgical therapy. 12th ed. Philadelphia: Elsevier; 2017. p. 360–5.

[2] Fong ZV, Tanabe KK. Ablation of colorectal carcinoma liver metastases. In: Cameron JL, Cameron AM, editors. Current surgical therapy. 12th ed. Philadelphia: Elsevier; 2017. p. 371–5.

[3] Medina BD, DiMatteo RP. Metastatic colorectal carcinoma. In: McIntyre Jr RC, Schulick RD, editors. Surgical decision making. 6th ed. Philadelphia: Elsevier; 2020. p. 292–4.

CHAPTER 12

Hilar, bile duct cancer, and cholangiocarcinoma

- Cholangiocarcinoma comprises 3% of all gastrointestinal malignancies. It most frequently occurs in the 6th–7th decade of life. The incidence of cholangiocarcinoma is approximately 0.5–1.5/100,000 in Western countries. It is prevalent in Southeast Asian countries; in Thailand the incidence is 115/100,000.
- Risk factors associated with cholangiocarcinoma include sclerosing cholangitis, congenital lesions such as choledochal cysts, parasitic infections with *Clonorchis sinensis* and *Opisthorchis viverrini*, as well as hepatitis B/C, and *Salmonella* infection.
- The distribution and frequency of cholangiocarcinoma is as follows:
  - Intrahepatic (5%–10%)
  - Hilar (60%–70%), also known as Klatskin tumors
  - Distal (20%–30%)
- Patients may present with abdominal pain, weight loss, nausea, anorexia, and jaundice (hilar and distal cholangiocarcinoma). Laboratory values of bilirubin, AST/ALT, alkaline phosphatase, and gamma-glutamyl transferase may be elevated.
- Transabdominal ultrasound may reveal dilated intrahepatic ducts with a normal common bile duct (hilar cholangiocarcinoma), dilated intra/extrahepatic ducts (distal cholangiocarcinoma), or a hepatic mass (intrahepatic cholangiocarcinoma).
- CT/MRI-MRCP is useful for tumor localization, and may depict bile duct stricture, liver atrophy, vascular involvement, and metastatic disease. These modalities are important in assessing lesion operability.
- Diagnosis of intrahepatic cholangiocarcinoma may be challenging. A primary extrahepatic tumor (e.g., colorectal cancer) must be ruled out. On CT scan, intrahepatic cholangiocarcinoma may be seen as mass with low attenuation, minor peripheral enhancement, and upstream biliary dilatation, suggesting biliary obstruction.

Digestive System Malignancies
https://doi.org/10.1016/B978-0-323-98369-3.00005-8

- ERCP may reveal the bile duct stricture and may provide tissue for histology. Drainage/stenting of the biliary tree is also possible.
- Up to 50% of patients with cholangiocarcinoma have unresectable disease at presentation.
- Several staging systems exist for hilar cholangiocarcinoma. The Bismuth-Corlette system considers the location of tumor with regard to location within the biliary tree, as well as the relationship of tumor with the confluence of (and extension into) the major hepatic ducts. The Blumgart system additionally examines the presence or absence of liver atrophy as well as portal venous involvement. This system is optimal for preoperative planning. The TNM system is usesful for postoperative risk stratification.
- In general, unresectable disease includes tumors with bilateral involvement of the hepatic ducts up to secondary biliary radicles, portal venous encasement proximal to its bifurcation.
- Surgery for hilar cholangiocarcinoma involves resection of the involved biliary tract; an extensive hepatectomy with Roux-en-Y hepaticojejunostomy is frequently performed in order to achieve an R0 resection.
- Surgery for intrahepatic cholangiocarcinoma usually requires hepatectomy with hilar lymph node dissection.
- Surgery for distal cholangiocarcinoma usually requires pancreaticoduodenectomy.
- Multiple lesions and/or recurrences may be treated with TAE/TACE, or radiofrequency ablation.
- Adjuvant chemotherapy with capecitabine in resected patients has been associated with increased overall survival. Gemcitabine + cisplatin are alternative chemotherapeutic agents.
- Immunotherapy (pembrolizumab-a programmed death 1 inhibitor) can be used in unresectable or metastatic refractory cholangiocarcinoma. High microsatellite instability or mismatch-repair deficiency of the tumor is a prerequisite.
- Molecularly targeted therapy has been recently added to the armamentarium of treating refractory cholangiocarcinoma. Pemigatinib, an oral small-molecule inhibitor of fibroblast growth factor receptor (FGFR), has become the first drug approved by the FDA specifically for the treatment of cholangiocarcinoma.

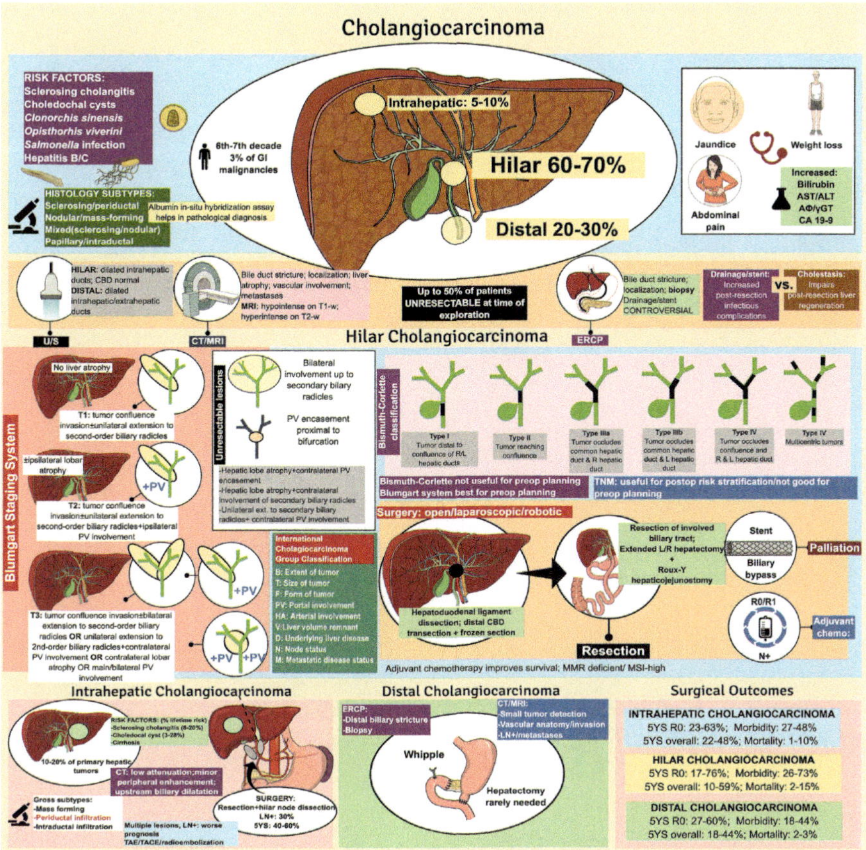

Fig. 1 Hilar and distal bile duct carcinoma: Classification, staging, and treatment.

Further reading

[1] Schulick RD, Gleisner AL. The management of malignant liver tumors. In: Cameron JL, Cameron AM, editors. Current surgical therapy. 12th ed. Philadelphia: Elsevier; 2017. p. 360–5.

[2] Gholami S, Jarnagin WR. Cholangiocarcinoma. In: McIntyre Jr RC, Schulick RD, editors. Surgical decision making. 6th ed. Philadelphia: Elsevier; 2020. p. 244–6.

[3] Nagorney DM, Pawlik TM, Chun YS, et al. Perihilar bile ducts. In: Amin MB, editor. AJCC cancer staging manual. 8th ed. Chicago: AJCC; 2017. p. 311.

[4] Ebata T, Kosuge T, Hirano S, et al. Proposal to modify the International Union Against Cancer staging system for perihilar cholangiocarcinomas. Br J Surg 2014;101:79.

CHAPTER 13

Gallbladder cancer

- Gallbladder cancer is rare; the incidence is approximately 1.4/100,000 in females and 0.8/100,000 in males. In the United States, 5000 cases are reported annually, with 2800 deaths per year. Gallbladder cancer is found in 1% of cholecystectomy specimens.
- Risk factors include the chronic presence of large gallstones, chronic cholecystitis, porcelain gallbladder (cancer risk 2%–3%), large gallbladder polyps (> 1 cm), female sex, Alaskan American or Native American heritage, and obesity.
- Gallbladder cancer may present in the following manner:
  - Preoperatively on workup for biliary symptoms.
  - As an incidental finding on imaging.
  - Intraoperatively during cholecystectomy.
  - Postoperatively in the pathology specimen.
- Approximately 50% of gallbladder cancers are diagnosed preoperatively; most tumors are advanced at presentation (37% are stage IV gallbladder cancer).
- Gallbladder cancer is in the fundus (60% of cases), the body (30% of cases), and the neck in 10% of cases.
- Two distinct histological patterns are found in gallbladder cancer. The diffusely nodular patter spreads along the subserosal plane and has better prognosis. The nodular pattern is characterized by the early invasion of the gallbladder wall and has worse prognosis.
- In most cases of gallbladder cancer, clinical presentation is similar to benign cholecystopathy. Advanced disease may present with jaundice and weight loss with or without abdominal pain.
- Abdominal ultrasound, CT, MRI/MRCP are the most common modalities used in the diagnosis of gallbladder cancer. Hepatic involvement, biliary extension, vascular involvement, ascites, and metastatic disease can be depicted. Lymph node involvement of the hepatoduodenal ligament can be demonstrated with high-quality imaging.
- Of note, tumors in Hartmann's pouch may resemble hilar cholangiocarcinoma.

Digestive System Malignancies
https://doi.org/10.1016/B978-0-323-98369-3.00008-3

- An elevated CEA (> 4 ng/mL) and/or CA 19-9 (> 200 U/mL) combined with suspicious findings on imaging may increase diagnostic sensitivity and specificity.
- Gallbladder cancer staging dictates resectability as well as extent of operative resection. Thus, careful pathological examination of gallbladder specimens, and high-quality imaging (if incidental gallbladder carcinoma is found) is important in planning for treatment.
- T1a disease is characterized by tumor invasion of the lamina propria; in T1b disease, tumor invades the gallbladder muscle layer. This distinction is extremely important.
- N1 disease includes metastases in the cystic duct, common bile duct, portal venous, and hepatic artery nodes. N2 disease includes metastases in celiac axis, periaortic, pericaval, and superior mesenteric artery nodes.
- Peritoneal carcinomatosis is common in gallbladder cancer (23% of cases). This may not be diagnosed using imaging techniques; in many centers, laparoscopy is utilized prior to planned resection.
- Cholecystectomy is adequate treatment in patients with T1a disease. The cystic duct margin must be examined and should be negative. In cases of positive margins, the common hepatic duct, common bile dust must be resected and a Roux-en-Y hepaticojejunostomy performed.
- In cases of T1b disease, lymph node metastases develop in 11% of patients. Therefore, radical cholecystectomy is recommended. This procedure includes segment IV/V resection, portal lymph node dissection, extrahepatic duct resection with Roux-en-Y reconstruction.
- T3/T4 lesions are advanced and have poor prognosis. In carefully selected cases, major hepatic resection may be performed. Operative morbidity and mortality is notoriously high.
- Unresectable disease includes patients with distant metastases, peritoneal surface malignancy, malignant ascites, and vessels encased in tumor.
- Adjuvant chemoradiotherapy may offer a survival benefit in patients having undergone resection. 5-FU and gemcitabine are the most common drugs used.
- Adjuvant chemoradiotherapy may offer modest survival benefit in patients with unresectable disease. Biliary stents/percutaneous biliary tree drainage may also be employed. The overall survival in gallbladder cancer patients is low. Stage I patients have a 5-year survival of 39%–43% with best-practice treatment. Patients with more advanced lesions have poor outcomes.

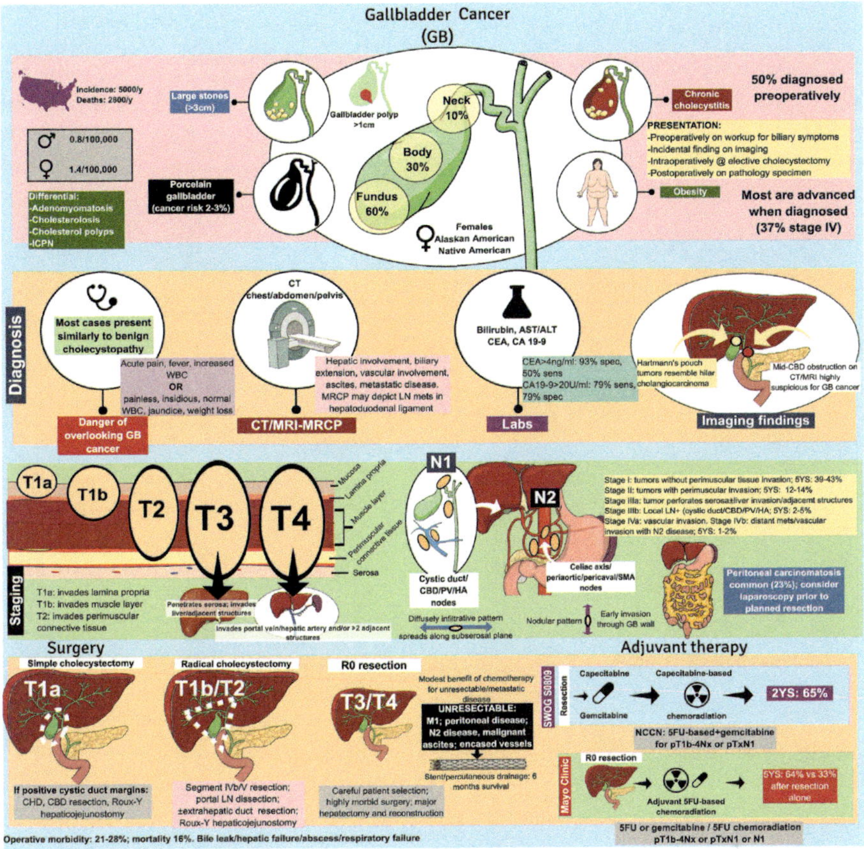

Fig. 1 Gallbladder cancer.

Further reading

[1] Marmor RA, Sicklick JK. The management of gallbladder cancer. In: Cameron JL, Cameron AM, editors. Current surgical therapy. 12th ed. Philadelphia: Elsevier; 2017. p. 466–72.

[2] Lewis A, Fong Y. Gallbladder carcinoma. In: McIntyre Jr RC, Schulick RD, editors. Surgical decision making. 6th ed. Philadelphia: Elsevier; 2020. p. 255.

CHAPTER 14
Appendiceal neoplasms

- Appendiceal neoplasms are rare lesions comprising 0.5%–1% of intestinal neoplasms. Most are incidental postappendectomy pathological findings. They may be found in 1% of appendectomy specimens. Epithelial neoplasms and neuroendocrine tumors (NET) comprise the majority of appendiceal tumors.
- **Epithelial neoplasms** are currently the most common appendiceal neoplasms. They are divided into:
  - Low-grade appendiceal mucinous neoplasms (LAMN)
  - High-grade appendiceal mucinous neoplasms (HAMN)
  - Invasive adenocarcinomas
  - Goblet-cell adenocarcinomas (adenocarcinoma + neuroendocrine tumor)
- A mucocele is a nonneoplastic appendiceal lesion filled with mucin.
- Epithelial neoplasms may rupture, leading to complicated clinicopathological conditions.
- Pseudomyxoma peritonei (PMP) or diffuse peritoneal adenomucinosis (DPAM) is the result of a ruptured mucocele or LAMN. It consists of histologically benign peritoneal mucin with a variable extent of spread in the peritoneal cavity
- Peritoneal mucinous adenocarcinomatosis (PMAC) is the result of a ruptured mucinous adenocarcinoma. It consists of peritoneal mucin with a cellular, cribriform growth pattern containing cells with high-grade atypia.
- Rupture of epithelial appendiceal neoplasms results in indolent progressive growth of the peritoneally spread lesion (PMP/DPAM/PMAC). Patients present with increasing abdominal girth, ascites, and abdominal pain. The outcome may be fatal secondary to ileus, if left untreated.
- Epithelial appendiceal neoplasms are staged according to the TNM system. Of note, lesions with intraperitoneal acellular mucin (PMP/DPAM) are classified as M1 disease.
- Surgical resection of an intact mucocele is considered adequate treatment.
- Appendectomy alone may be considered adequate surgical treatment of a nonruptured appendiceal adenocarcinoma if it is well-differentiated,

Digestive System Malignancies
https://doi.org/10.1016/B978-0-323-98369-3.00016-2

with tumor invasion up to the submucosa. Stage I–III adenocarcinomas without rupture are treated with a right hemicolectomy. Adjuvant chemotherapy is recommended for T4 lesions, poorly differentiated tumors, inadequate lymph node retrieval.

- Stage IV adenocarcinoma (rupture) and goblet-cell adenocarcinoma may be treated with cytoreductive surgery and HIPEC.
- PMP/DPAM is treated with periodic surgical debulking or cytoreductive surgery with HIPEC. Optimal cytoreductive surgery is achieved when no residual disease remains, or residual deposits are smaller than 2–2.5 cm.
- **Neuroendocrine tumors** of the appendix are more common in the distal third of the appendix; 10% are located at the appendiceal base.
- Carcinoid syndrome is extremely rare.
- The WHO pathological classification of NETs considers histologic grade (proliferative tumor activity/mitotic rate and Ki67 index), and differentiation.
- Appendiceal NETs less than 1 cm in size, or lesions 1.1–1.9 cm and no mesoappendiceal invasion/positive or unclear margins are treated with appendectomy alone.
- Appendiceal lesions >1 cm or lesions 1.1–1.9 cm with mesoappendiceal invasion/positive or unclear margins are treated with right hemicolectomy.
- For tumors larger than 2 cm, staging with CT/MRI is recommended. Endoscopy may reveal synchronous GI lesions in 25% of patients.
- Appendiceal NETs rupture in 10%–20% of cases. Perforation has no impact on the treatment plan; peritoneal recurrence is rare.
- For patients with resected tumors <2 cm in size, no surveillance is recommended (NCCN). For tumors >2 cm, physical exam 3–12 months after resection is undertaken, with consideration of CT/MRI and chromogranin A assay biannually/annually for 7 years postresection. Similar surveillance may be offered to patients with resected lesions 1–2 cm in size with poor prognostic factors (node positive, lymphovascular invasion, mesoappendiceal invasion, intermediate, high-grade, or mixed histology).

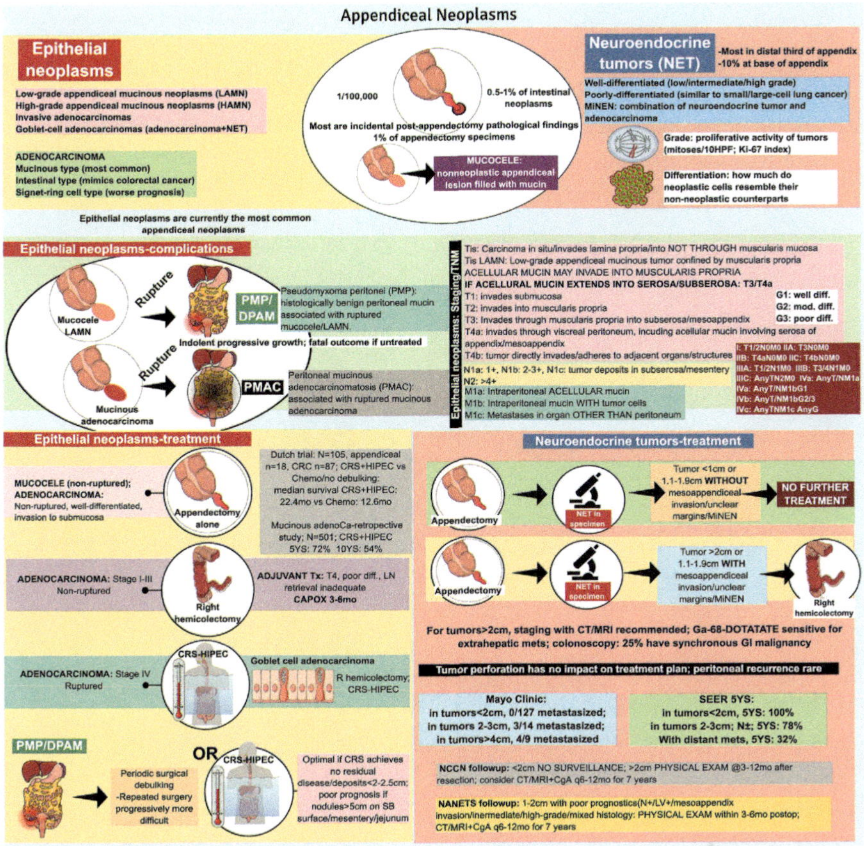

Fig. 1 Appendiceal neoplasms.

Further reading

[1] Spanos CP, Kaiser AM. Appendiceal neoplasms. In: Steele SR, Hull TR, Read TE, Saclarides TJ, Senagore AJ, Whitlow CB, editors. The ASCRS textbook of colon and rectal surgery. 3rd ed. Heidelberg: Springer; 2016. p. 617–30.

[2] Turaga KK, Pappas SG, Gamblin T. Importance of histologic subtype in the staging of appendiceal tumors. Ann Surg Oncol 2012;19:1379.

CHAPTER 15
Colon cancer

- Colon and rectal cancer (CRC) affects approximately 1 million individuals worldwide annually. In Western countries, disease-specific mortality has been declining by 1.6%–2% per year on account of screening programs. However, the incidence of colorectal cancer in individuals under 50 years of age has been increasing by 2.1% per year. Males and females are affected equally.
- The distribution of colon and rectal cancer is as follows:
    - Cecum: 12%–17%
    - Ascending colon: 7%–10%
    - Hepatic flexure: 3%
    - Transverse colon: 5%–6%
    - Splenic flexure: 2%
    - Descending colon: 3%
    - Sigmoid colon: 23%–30%
    - Rectum: 23%–32%
- Most colorectal cancer is sporadic and is derived from adenomas. Chronic inflammatory bowel disease (ulcerative colitis, Crohn's disease) increases the risk for developing CRC.
- Inherited forms of CRC occur in 3%–4% of patients. Hereditary nonpolyposis colorectal cancer is the result of germline DNA mismatch-repair genes. Familial adenomatous polyposis (FAP) is the result of a germline mutation of the APC gene.
- Adenomatous polyps form in the colon when normal mechanisms regulating epithelial renewal are disrupted. Surface cells lining the intestine are continuously lost into the bowel lumen due to apoptosis and exfoliation and must be continuously replaced. Typically, proliferation occurs at the base of colonic crypts. As cells move toward the luminal surface, they cease proliferating and terminally differentiate. This process is disrupted as adenomas increase in size, become dysplastic, and eventually become invasive. This is known as the adenoma-carcinoma sequence. Mutations of several oncogenes and tumor-suppressor genes are involved in this sequence.

Digestive System Malignancies
https://doi.org/10.1016/B978-0-323-98369-3.00020-4

- Colon and rectal cancer may present:
  - On screening endoscopy in asymptomatic patients
  - With suspicious symptoms
  - Emergently with obstruction, perforation, or bleeding
- In right-sided colon cancer, suspicious symptoms include anemia, weakness, malaise, and abdominal pain. These are nonspecific symptoms. Lesions in the vicinity of the ileocecal valve may cause obstructive symptoms (abdominal pain, nausea, and vomiting).
- In left-sided CRC, suspicious symptoms include changes in bowel habits, constipation, blood per rectum, abdominal pain, tenesmus, and obstruction.
- Unusual presentations of CRC include colovesical fistula, colovaginal fistula, coloenteric fistula, *Streptococcus bovis* bacteremia, and *Clostridium septicum* sepsis.
- Colonoscopy is the mainstay of CRC diagnosis. Biopsies are obtained. It may detect synchronous cancer in 3%–5% of cases and synchronous polyps in 50%.
- Stool DNA testing is currently available for CRC detection.
- CT colonography may be utilized as the initial diagnostic test or if colonoscopy is incomplete. If a lesion is detected, endoscopy with biopsy is necessary.
- If histopathologic diagnosis of CRC is made after endoscopy, a CT of the chest/abdomen and pelvis is performed. This may reveal the extent of regional tumor involvement, depict lymph node suspicious for metastatic disease as well as distant metastases. Peritoneal surface malignancy is notoriously elusive regarding imaging with CT-scan.
- Approximately 20% of patients with CRC have distant metastases on initial presentation.
- CEA (carcinoembryonic antigen) is the primary tumor marker used in CRC. Of note, it has low diagnostic ability for primary CRC diagnosis. Rising CEA levels after treatment imply local or distant recurrence.
- PET/CT is used for posttreatment rising CEA when conventional imaging is nondiagnostic.
- The TNM staging system is commonly used for CRC. The Dukes system is equally useful.
- Molecular profiling in CRC is extensively used for cancer prognosis and treatment planning. Microsatellite instability (MRI) is found in right sided tumors, portends poor differentiation, mucinous tumors and predicts a low response to 5FU-based adjuvant chemotherapy. Tumors with K-ras/N-ras/BRAF mutations predict lack of response to EGFR-targeting agents, such as cetuximab.

- Surgery is the mainstay of colon cancer treatment. Principles of surgical resection include:
    - Adequate resection luminal margins (5–7 cm from tumor edge)
    - Adequate mesenteric/mesocolic lymphadenectomy (at least 12 lymph nodes in specimen)
    - High ligation of feeding vessels
    - Adequate blood supply and lack of tension to ensure a safe intestinal anastomosis
    - In general, a D2 lymphadenectomy should be performed. This includes regional mesenteric lymph nodes along the course of major mesenteric vessels and vascular arcades
- The **lymph node ratio** (number of positive lymph nodes divided by total harvested lymph nodes) is a significant prognostic factor.
- **Complete mesocolic excision** refers to anatomic resection of the mesentery with an intact visceral peritoneum, tumor-free bowel proximal and distal to tumor, and central venous ligation, thus retrieving lymph nodes and tissue overlying the superior mesenteric vein and artery (D3 lymph nodes).
- Conventional right hemicolectomy includes ligation of the ileocolic artery, right colic artery, and the right branch of the middle colic artery.
- Extend right hemicolectomy refers to ligation of the ileocolic artery, right colic artery, and the middle colic artery.
- In a left hemicolectomy the left branch of the middle colic and left colic are ligated.
- During a sigmoid resection the inferior mesenteric and superior rectal artery are ligated.

  In patients with synchronous tumors, HNPCC, MUTYH mutation and left-sided obstruction, a total abdominal colectomy with an ileorectal anastomosis.
- In patients with FAP, proctocolectomy with construction of an ileal pouch-anal anastomosis is performed.
- Surgery for colorectal cancer may be performed with minimally invasive techniques (laparoscopic, robotic). Oncologic outcomes have been found to be equivalent to open techniques with prospective, randomized trials.
- Adjuvant chemotherapy is recommended in patients with lymph node metastases. Six months of FOLFOX is a commonly administered regimen. Tumor molecular profiling may aid in choice of drugs used for treatment.

- Adjuvant chemotherapy may also be recommended in "high-risk N0" disease. This may include:
  - Poorly differentiated tumors
  - Tumors with lymphovascular invasion
  - Perforated tumors
  - Resections with inadequate margins
  - T4 lesions
  - Specimens with less than 12 lymph nodes retrieved

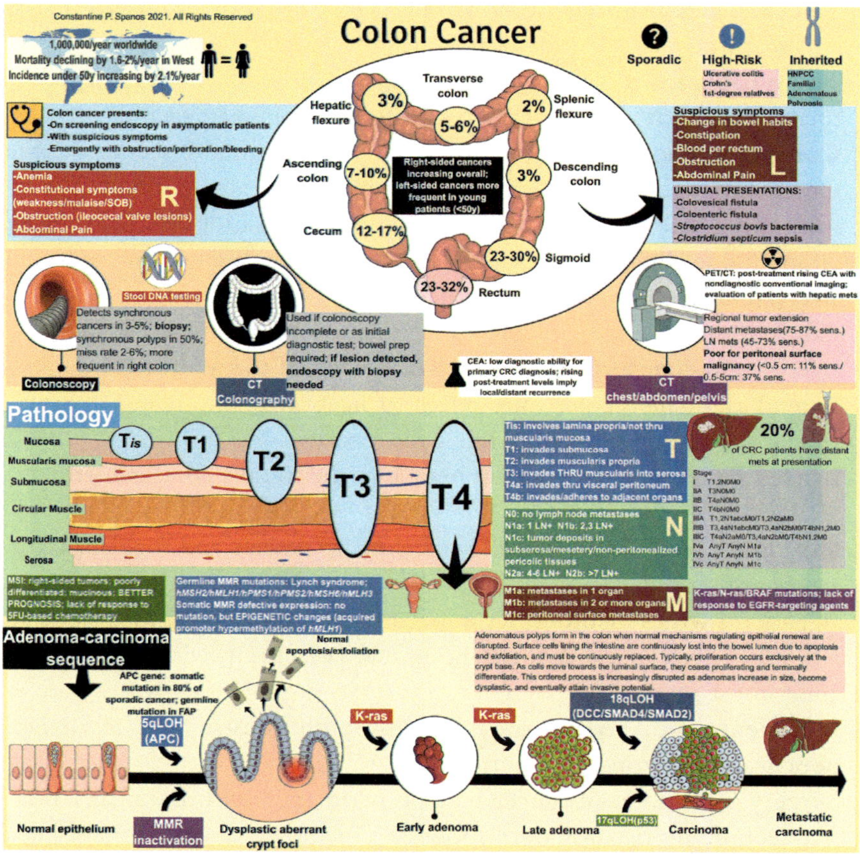

Fig. 1 Colon cancer: Classification, evaluation, adenoma-carcinoma sequence.

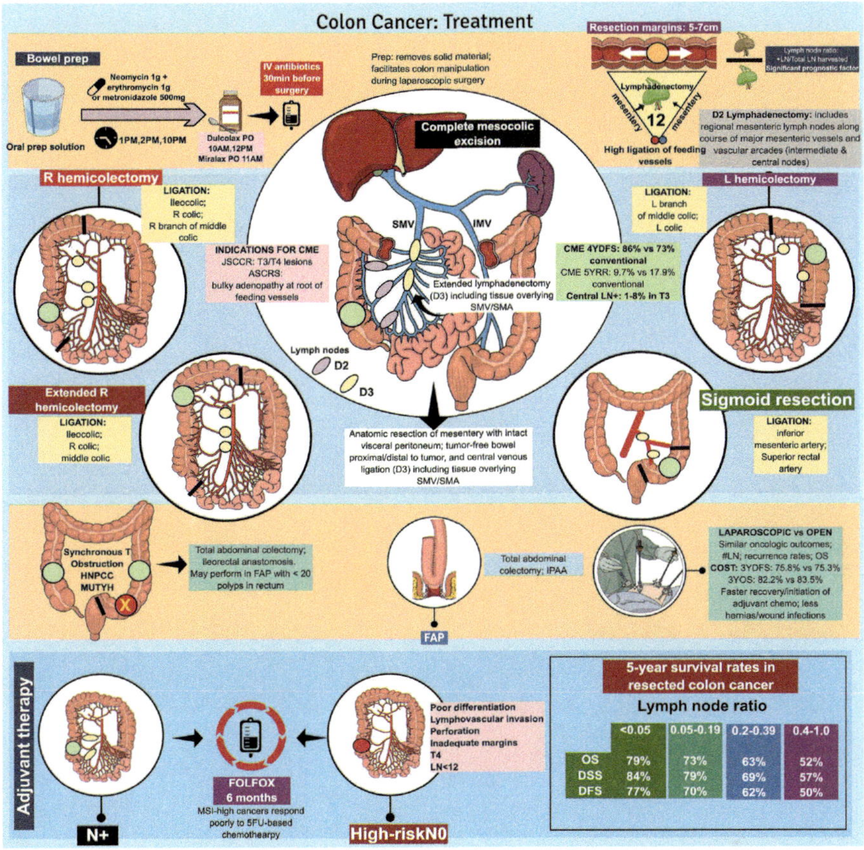

Fig. 2 Colon cancer: Treatment.

Further reading

[1] Chang GJ, Kaiser AM, Mills S, et al. Practice parameters for the management of colon cancer. Dis Colon Rectum 2012;55:831.

[2] Bertelsen CA, Neuenschwander AU, Jansen JE, et al. Short-term outcomes after complete mesocolic excision compared with 'conventional' colonic cancer surgery. Br J Surg 2016;103:581.

[3] Anandam JL, Choti MA. The management of colon cancer. In: Cameron JL, Cameron AM, editors. Current surgical therapy. 12th ed. Philadelphia: Elsevier; 2017. p. 222–4.

CHAPTER 16

Rectal cancer

- Rectal cancer affects approximately 80,000 individuals in the European Union annually. Most colorectal cancer is sporadic and is derived from adenomas. Chronic inflammatory bowel disease (ulcerative colitis, Crohn's disease) increases the risk for developing CRC.
- Inherited forms of CRC occur in 3%–4% of patients. Hereditary non-polyposis colorectal cancer is the result of germline DNA mismatch-repair genes. Familial adenomatous polyposis (FAP) is the result of a germline mutation of the APC gene.
- Rectal cancer is frequently symptomatic and may present with blood in rectum, a change in bowel habits, tenesmus, and rarely rectal pain.
- A digital rectal exam may reveal a palpable mass. In addition, it may reveal important information such as size, lesion circumference relative to the lumen, whether the lesion is anterior or posterior, fixed, or mobile, its relationship to the sphincter mechanism, and sphincter function.
- Rigid or flexible proctoscopy may reveal lesion distance from the anal verge. Colonoscopy is useful to rule out synchronous lesions and pathology (e.g., ulcerative colitis).
- Pretreatment assessment is usually performed to clinically "stage" rectal cancer. The most common modalities used are endorectal ultrasound, MRI of the abdomen and pelvis and CT scan of the chest abdomen and pelvis. This assessment is fundamental in planning for rectal cancer treatment.
- Endorectal ultrasound has a relatively high sensitivity for tumor depth of invasion assessment (T stage). It is not as accurate in detecting possible lymph node metastases.
- MRI is relatively accurate for detecting transmural tumor invasion, possible lymph node metastases, extramural lymph node invasion (EMVI), as well as predicting the circumferential resection margin (CRM). The CRM is the distance from the maximum depth of tumor and the pelvic sidewall; a distance < 1 mm is predictive of local recurrence after resection.
- CT scan is not accurate in the assessment of tumor depth of invasion. It may depict invasion of adjacent organs and structures, as well as distant metastases.
- The TNM system is commonly used for staging.

Digestive System Malignancies
https://doi.org/10.1016/B978-0-323-98369-3.00018-6

- The surgical standard of care for rectal cancer is total mesorectal excision (TME). TME employs a precise, sharp dissection between the visceral and parietal layers of the endopelvic fascia to ensure en-bloc removal of the perirectal areolar tissue, including the lateral and circumferential margins of the mesorectal envelope, lymphatics, and vascular/perineural tumor deposits with the primary rectal cancer. TME also preserves the autonomic nerves and reduces the risk of presacral bleeding.
- Sphincter-saving procedures include low anterior resection with colorectal anastomosis, ultralow anterior resection with coloanal anastomosis, and intersphincteric resection with coloanal anastomosis. Transanal total mesorectal excision (TaTME) is a novel sphincter-preserving procedure used in rectal cancer treatment. A temporary diverting ostomy may be fashioned.
- Abdominoperineal resection may be performed for low-lying rectal tumors involving the sphincter mechanism. The rectum and sphincters are resected and an end-colostomy is fashioned.
- Clinical stage I patients proceed directly to surgery.
- Clinical stage II, III patients usually undergo neoadjuvant chemoradiotherapy.
- Long-course neoadjuvant therapy involves combined modality with radiosensitizing chemotherapy (5-FU/LV or capecitabine) and concurrent radiotherapy (50.4 Gy total radiation dose over the course of 5–6 weeks) with surgical resection (TME) 4–8 weeks after completion of chemoradiotherapy. This is based largely on the German Rectal Cancer Trial, which randomized patients with T3–T4 or node-positive tumors to receive either preoperative or postoperative chemoradiation. All patients underwent TME followed by four additional cycles of 5-FU. Preoperative chemoradiation was associated with a significantly lower local recurrence (6% vs 13% with postoperative therapy).
- Short-course radiotherapy (25 Gy given in 5 fractions) followed by TME is used in some institutions. This approach has demonstrated an improved local disease control in clinical trials.
- Total neoadjuvant therapy (i.e., oxaliplatin-based chemotherapy combined with long-course CRT or short-course RT, followed by TME) in patients with locally advanced rectal cancer who are at high risk for a margin-positive resection as well as for those with clearly node-positive disease and a low-lying rectal tumor, has recently been adopted as a treatment option in some centers.
- Local excision of rectal cancer may be a treatment option in highly selective cases (usually clinical T1N0). These include:
  - Tumor diameter < 3 cm
  - Tumor circumference < 1/3 of lumen

- Mobile, nonfixed tumor
- Achievable clear margins
- Favorable histology (well or moderately differentiated)
- Absence of EMVI
- No evidence of lymph node metastases

- Transanal excision can be achieved with transanal endoscopic microsurgery (TEM), transanal minimally invasive surgery (TAMIS), and endoscopic submucosal dissection (ESD).
- Complete clinical response may occur in rectal patients after neoadjuvant chemoradiotherapy. Patients with an apparent clinical complete response to neoadjuvant therapy should typically be offered radical resection, although a "watch and wait" management approach can be considered for highly selected patients in the context of a protocolized setting.

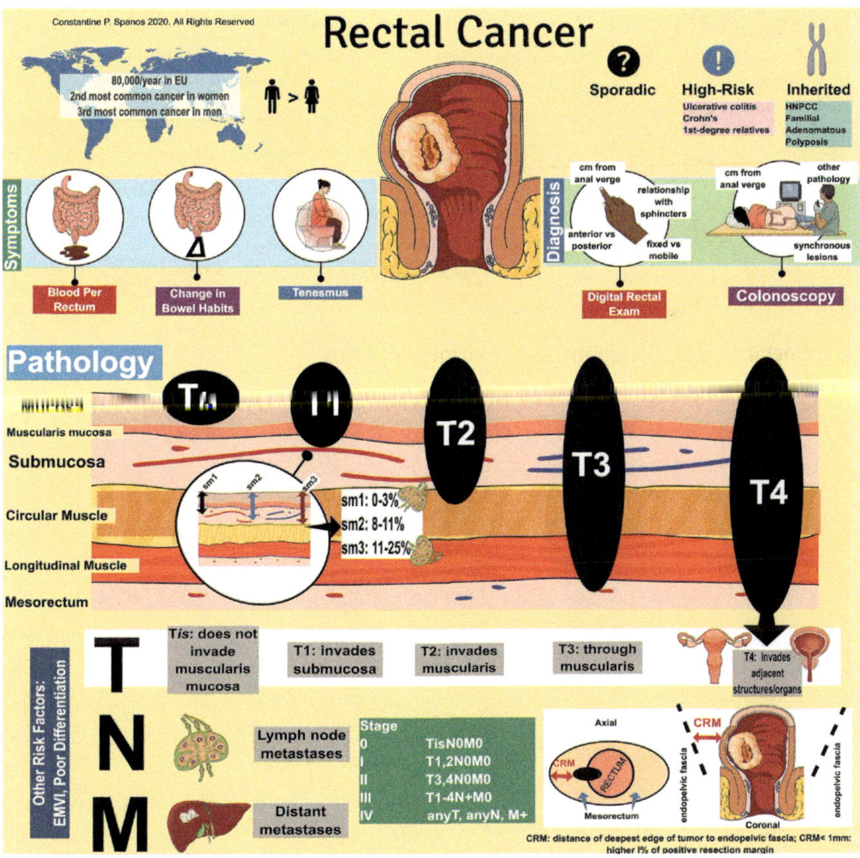

Fig. 1 Rectal cancer: Symptoms and staging.

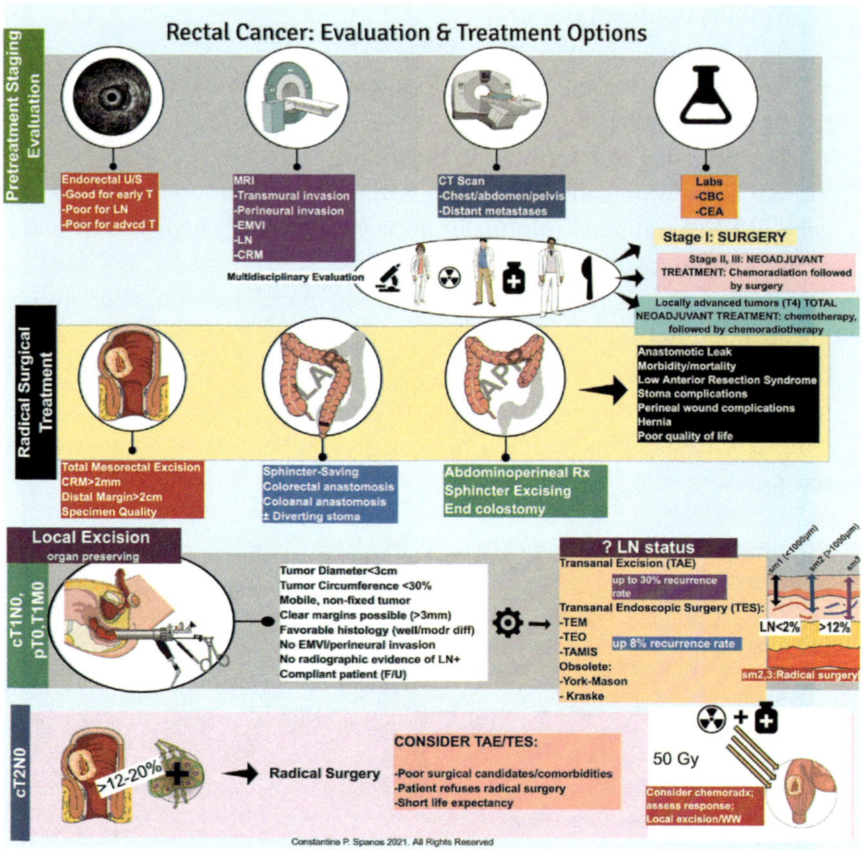

Fig. 2 Rectal cancer: Evaluation and treatment options.

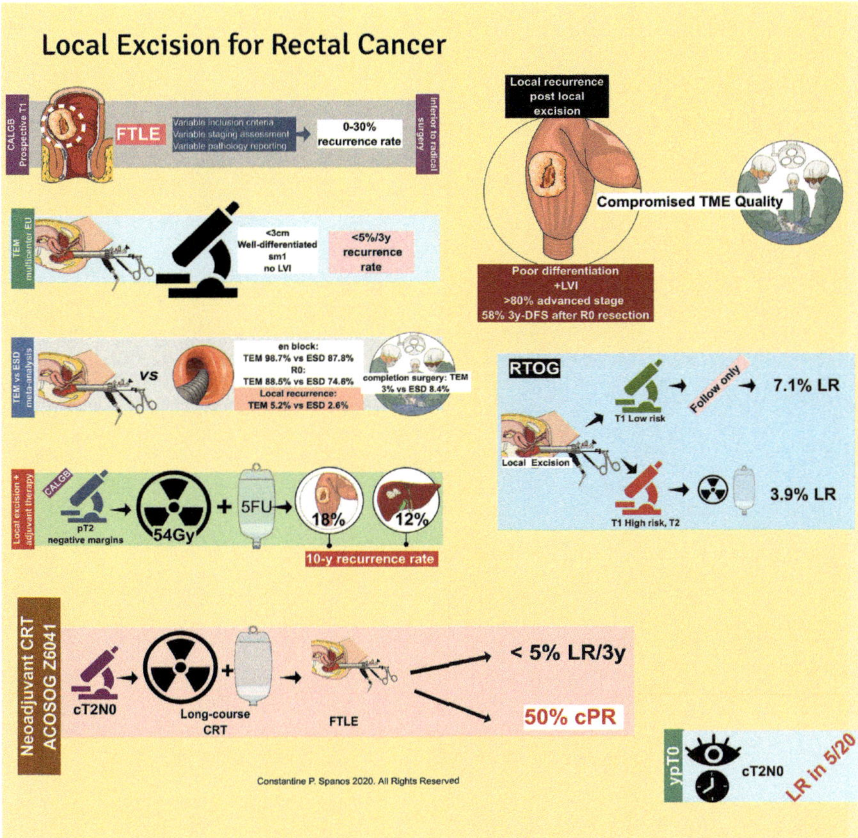

Fig. 3 Rectal cancer: Local excision for rectal cancer.

Further reading

[1] Kleiman DA, Guillem JG. The management of rectal cancer. In: Cameron JL, Cameron AM, editors. Current surgical therapy. 12th ed. Philadelphia: Elsevier; 2017. p. 224–34.

[2] Cercek A, Garcia-Aguilar J. Rectal cancer: neoadjuvant therapy. In: Steele SR, Hull TR, Read TE, Saclarides TJ, Senagore AJ, Whitlow CB, editors. The ASCRS textbook of colon and rectal surgery. 3rd ed. Heidelberg: Springer; 2016. p. 481–94.

[3] Whiteford MH. Local excision of rectal neoplasia. In: Steele SR, Hull TR, Read TE, Saclarides TJ, Senagore AJ, Whitlow CB, editors. The ASCRS textbook of colon and rectal surgery. 3rd ed. Heidelberg: Springer; 2016. p. 495–506.

CHAPTER 17

Anal cancer

- Anal cancer comprises 2.7% of malignancies of the digestive tract. In the United States, approximately 9090 cases are reported annually. The incidence of anal cancer has been increasing.
- Anal cancer includes squamous cell cancer (SCC), adenocarcinoma (deriving from the anal crypts), neuroendocrine tumors, and melanoma. SCC is by far the most common anal cancer.
- Regarding site of origin, anal cancer can arise from:
  - The anal transition zone (nonkeratinizing SCC).
  - The anal canal (keratinizing/nonkeratinizing SCC).
  - The anal crypts (adenocarcinoma).
  - The hair-bearing skin distal to the mucocutaneous junction (keratinizing SCC).
- Eponymous cancers arising in the perianal skin include Bowen's disease, which is an in situ SCC, and Paget's disease, which is an intraepithelial adenocarcinoma.
- Risk factors for anal SCC include:
  - Sexual activity and sequelae (multiple sexual partners, genital warts, HSV2 infection, chlamydia, receptive anal intercourse, and cervical cancer)
  - HPV infection. Normal epithelium infected with HPV may undergo dysplastic changes, leading to squamous intraepithelial lesions (SIL). High-grade SIL (HSIL) is premalignant; progression to SCC may occur in HIV patients with low CD4 levels, as well as high levels of high-risk HPV serotypes (HPV 6,11,16,18).
  - HIV infection. The incidence of SCC in HIV+ individuals twice that of HIV− individuals in males, but no in females. The etiopathogenesis is unclear. Of note, active antiretroviral therapy has not decreased the incidence of SCC in patients with HIV
  - Patients on immunosuppressive therapy (e.g., patients with solid organ transplantation).

Digestive System Malignancies
https://doi.org/10.1016/B978-0-323-98369-3.00015-0

- Patients with anal cancer present with rectal bleeding, anal pain, and a mass. Delay in diagnosis is frequent; patients and healthcare providers may confuse anal cancer with benign anorectal disorders.
- Physical examination includes inspection, digital rectal exam, and groin examination for lymphadenopathy. Lymph node metastases are present in 37% of patients. A gynecological examination to rule out cervical pathology is important.
- Evaluation for HIV (including history of antiretroviral therapy, CD4 levels, and viral load) is undertaken. Additional patients are screened for viral hepatitis.
- CT scan of chest/abdomen/pelvis may detect distant metastases as well as local tumor extent. PET/CT is more sensitive for detection of node metastases.
- The TNM system is used for staging. **Tumor size**, rather than depth of invasion, is used in the T subclassification.
- Combined modality treatment (CMT) is the most common form of therapy for anal SCC. This includes radiation therapy (total dose: 54 Gy, in fractions of 2 Gy/day), 5-FU on days 1–4 and 29–32, and mitomycin C on day 1 and day 28. A complete pathological response may be achieved in 80% of patients.
- HIV+ patients may undergo combined modality treatment; however, complications of AIDS may decrease patients' tolerance to treatment.
- Careful assessment for approximately 8–12 weeks posttreatment will detect persistent or recurrent disease. Tumor regression may occur up to 26 weeks after combined modality therapy.
- Patients with persistent or recurrent disease may undergo abdominoperineal resection; 5-year survival rates are approximately 45%. Distant metastases develop in 10%–17% of patients after combined modality treatment. Chemotherapeutic or immunotherapeutic agents may be used for treatment.
- Local excision for SCC is acceptable if basement membrane invasion is <3 mm and horizontal spread of the tumor is <7 mm. Wide local excision is the treatment of choice for SCC of the perianal skin.

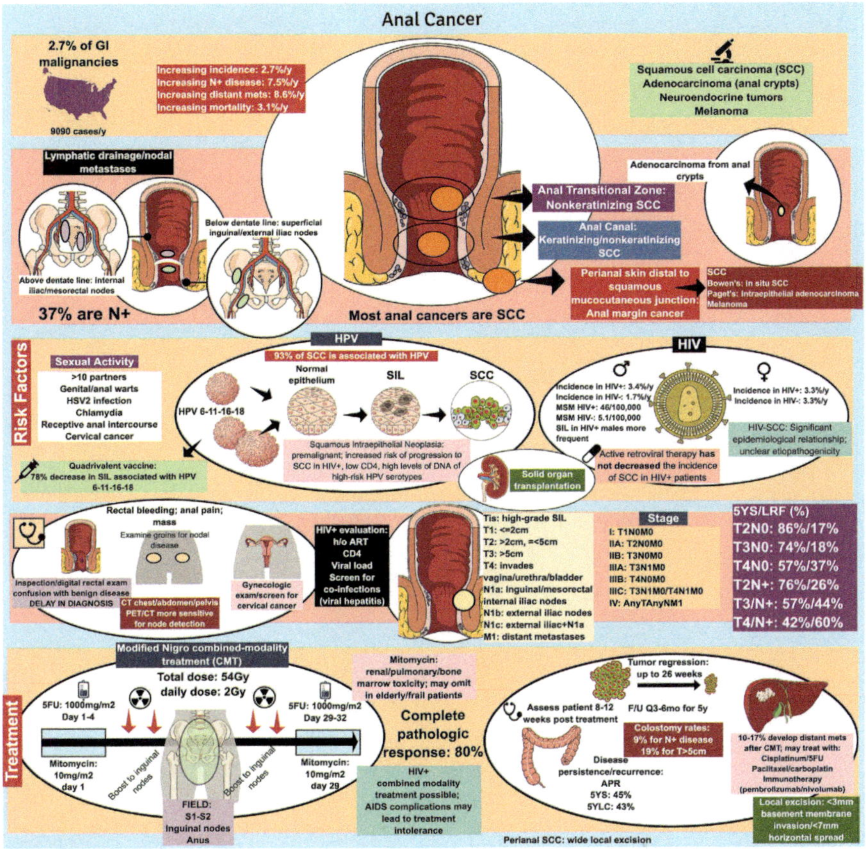

Fig. 1 Anal cancer.

Further reading

[1] Samdani T, Nash GM. Anal cancer. In: Steele SR, Hull TR, Read TE, Saclarides TJ, Senagore AJ, Whitlow CB, editors. The ASCRS textbook of colon and rectal surgery. 3rd ed. Heidelberg: Springer; 2016. p. 357–72.

[2] Shiels MS, Pfeiffer RM, Chaturvedi AK, et al. Impact of the HIV epidemic on the incidence rates of anal cancer in the United States. J Natl Cancer Inst 2012;104:1591.

[3] Islami F, Ferlay J, Lortet-Tieulent J, et al. International trends in anal cancer incidence rates. Int J Epidemiol 2016.

CHAPTER 18

Peritoneal surface malignancy

- Peritoneal surface malignancy (PSM) refers to the spread of cancer on the parietal peritoneum, the omentum, and the serosal surface of intraabdominal viscera. It is not considered a form of systemic distant metastasis. It is formerly known as peritoneal carcinomatosis.
- The distribution of PSM follows the "rule" of intraperitoneal fluid flow (clockwise toward the undersurface of the diaphragm), the phagocytic activity of the lesser and greater omentum and appendices epiploicae, and gravity.
- Lesions most commonly associated with PSM include:
  - Epithelial lesions of the appendix
  - Colorectal cancer
  - Ovarian cancer
  - Gastric cancer
  - Diffuse malignant peritoneal mesothelioma
- Epithelial neoplasms are currently the most common appendiceal neoplasms. They are divided into:
  - Low-grade appendiceal mucinous neoplasms (LAMN)
  - High-grade appendiceal mucinous neoplasms (HAMN)
  - Invasive adenocarcinomas
  - Goblet cell adenocarcinomas (adenocarcinoma + neuroendocrine tumor)
- A mucocele is a nonneoplastic appendiceal lesion filled with mucin.
- Epithelial neoplasms may rupture, leading to complicated clinicopathological conditions.
- Pseudomyxoma peritonei (PMP) or diffuse peritoneal adenomucinosis (DPAM) is the result of a ruptured mucocele or LAMN. It consists of histologically benign peritoneal mucin with a variable extent of spread in the peritoneal cavity.
- Peritoneal mucinous adenocarcinomatosis (PMAC) is the result of a ruptured mucinous adenocarcinoma. It consists of peritoneal mucin with a cellular, cribriform growth pattern containing cells with high-grade atypia.
- PSM may occur in 25% of cases of colorectal cancer, and 53%–60% of cases of gastric cancer.

Digestive System Malignancies
https://doi.org/10.1016/B978-0-323-98369-3.00009-5

- Patients with PSM may present with increasing abdominal girth, ascites, and hernia (umbilical, inguinal).
- CT/MRI may depict a large amount of intraabdominal mucinous material, scalloping of the liver, spleen and mesentery, or large cystic masses on the undersurface of the diaphragm. Another finding is the redistribution phenomenon, in which the small bowel is centrally displaced by the mucinous fluid.
- In many cases, PSM is found serendipitously during planned surgery (e.g., colorectal resection for cancer), or urgent surgery (e.g., exploration for bowel obstruction or appendicitis).
- Quantitative assessment for tumor distribution throughout the peritoneal cavity is performed using the **peritoneal carcinomatosis index (PCI)**, as described by Sugarbaker. The PCI is derived by examination of 13 abdominopelvic regions; a score is assigned to lesions of a certain size for each region (score 0–3). The cumulative score is the PCI. PCI is predictive of the adequacy of cytoreductive surgery. Preoperative PCI assessment is not as accurate as intraoperative assessment; combination of CT and MRI may increase sensitivity preoperatively.
- Operative treatment of PSM is comprised of cytoreductive surgery (CRS) and hyperthermic intraperitoneal chemotherapy (HIPEC).
- CRS consists of:
    - Greater omentectomy with splenectomy
    - Lesser omentectomy, hepatoduodenal dissection, and cholecystectomy
    - Rectosigmoid resection
    - Small bowel lesion resection
    - Pelvic peritonectomy, abdominal hysterectomy, and salpingoophorectomy
    - Total anterior parietal peritonectomy
    - Retroperitoneal peritonectomy
    - Upper quadrant peritonectomy (diaphragm)
- HIPEC is performed prior to any gastrointestinal anastomosis and consists of a chemotherapeutic peritoneal "bath" with mitomycin and cisplatinum. The solution is heated to 41°C, and the duration of the "bath" is 90 minutes. HIPEC increases drug penetration and toxicity, leads to impaired tumor DNA repair, protein denaturation, and cell death.
- High-volume centers offer acceptable 5-year survival rates for PSM associated with appendiceal lesions, colorectal cancer, diffuse malignant peritoneal mesothelioma, and ovarian cancer. A PCI <15 is predictive of adequate cytoreduction and better results.
- CRS + HIPEC is associated with significant morbidity (up to 65%) and mortality (up to 8%).

- Early postoperative intraperitoneal chemotherapy (EPIC), bidirectional chemotherapy (intraoperative and oral chemotherapy), as well as adjuvant chemotherapy may be offered to patients postoperatively.

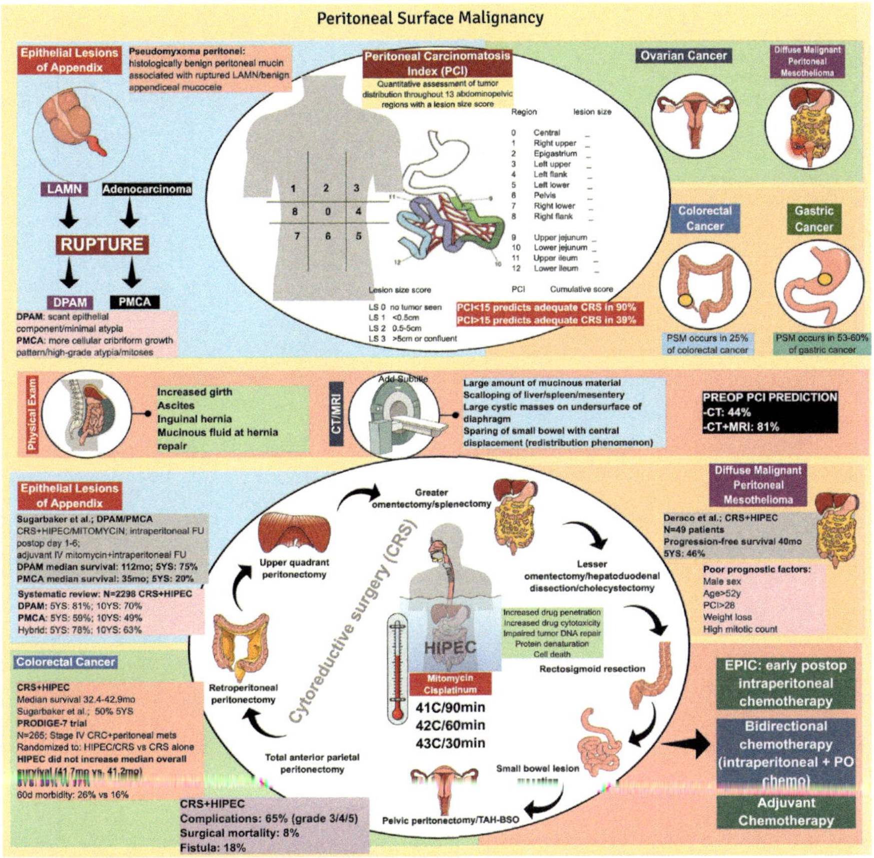

Fig. 1 Peritoneal surface malignancy.

Further reading

[1] Sugarbaker PH, Welch LS, Mohamed F, Glehen O. A review of peritoneal mesothelioma at the Washington Cancer Institute. Surg Oncol Clin N Am 2003;12:605.

[2] Gusani NJ, Cho SW, Colovos C, et al. Aggressive surgical management of peritoneal carcinomatosis with low mortality in a high-volume tertiary cancer center. Ann Surg Oncol 2008;15:754.

[3] Esquivel J. The management of peritoneal surface malignancies of appendiceal or colorectal origin. In: Cameron JL, Cameron AM, editors. Current surgical therapy. 12th ed. Philadelphia: Elsevier; 2017. p. 258–65.

[4] Canbay E, Yonemura Y. Anatomy of peritoneum and preoperative assessment of patients with peritoneal surface malignancies. In: Canbay E, Yonemura Y, editors. Peritoneal surface malignancies. A curative approach. Heidelberg: Springer; 2015. p. 1–19.

CHAPTER 19

Inflammation and digestive malignancies

- There is a well-established link between inflammation and carcinogenesis. Inflammation is associated with tissue injury and arises via innate immunity. The innate immune system includes macrophages, mast cells dendritic cells, granulocytes, and innate lymphocytes (NK cells, T cells).
- Exposure of tissue to physical injury, ischemia, toxins, and infectious agents may induce inflammation. This may provoke antitumor immune responses or promote tumor development.
- Approximately 15%–20% of cancers are attributable or associated with chronic inflammation secondary to infectious agents, chemical and physical agents, as well as autoimmune reactions.
- Inflammation is an integral component in all three stages of carcinogenesis:
  - Cancer initiation
  - Cancer promotion
  - Cancer progression
- During **cancer initiation**, genetic mutations lead to production of proteins with altered tertiary structure and dysregulation of normal biologic function. Initiating events include DNA alterations and hereditary mutations. Chronic inflammation leads to persistence of such initiating factors. The inflammatory microenvironment leads to increased mutation rates of cells, thus enhancing proliferation of mutated cells, production of reactive oxygen species (ROS), furthering DNA damage, and genomic instability. TNFα produced by inflammatory cells also leads to ROS production and oxidative cell damage.
- **Cancer promotion** is the process of tumor growth from single initiated cells to a fully developed primary tumor. Cytokines produced by inflammatory cells lead to activation of key transcription factors (NF-κβ, STAT3). "Initiated" cancer cells that are exposed to chemicals, factors released at sites of tissue damage, hormones, and tumor promoters, may additionally recruit inflammatory cells. These may cause oxidative DNA damage and reduce DNA repair. Ultimately, increased cell proliferation and decreased cell-death ensue.

Digestive System Malignancies
https://doi.org/10.1016/B978-0-323-98369-3.00014-9

- **Cancer progression** leads to cancer cell proliferative self-sufficiency, insensitivity to antigrowth factors, resistance to apoptosis, limitless replication potential, angiogenesis, invasion, and metastasis. These are the hallmarks of cancer.

Esophageal cancer and inflammation

- Esophageal cancer frequently arises from Barrett's esophagus. In Barrett's esophagus, the nonstratified squamous epithelium is replaced by mucin-secreting columnar-lined intestinal-type epithelium.
- Esophageal carcinogenesis is based on the metaplasia-dysplasia-adenocarcinoma sequence.
- Acid and bile reflux cause chronic inflammation in the squamous epithelium and esophageal glands.
- Metaplasia of the squamous epithelium ensues.
- Chronic inflammation induces the production of TNFα, NF-κβ, and cytokines. Ultimately cell damage caused by these processes leads to dysplasia.
- Further DNA damage and aneuploidy promotes cell proliferation and inhibition of apoptosis, leading to esophageal adenocarcinoma.

Inflammation and gastric cancer

- *Helicobacter pylori* is closely linked to gastric carcinogenesis. It is considered a class I carcinogen.
- *H. pylori* is associated with distal gastric tumors of the intestinal histological type.
- This gram-negative, microaerophilic bacterium causes chronic gastritis. A host immune response to infection leads to recruitment of neutrophils, lymphocytes, macrophages, which are activated.
- *H. pylori* virulence factors and host factors (virulence-associated gene polymorphisms) act synergistically in promoting inflammation, oxidative stress, gastric immune cell accumulation and proinflammatory cytokine production and cellular transformation.
- *H pylori*-associated atrophic gastritis is a key step in gastric carcinogenesis. The resulting gastric mucosal expression of cytokines is associated with inhibition of gastric acid secretion. This evolves into multifocal mucosal atrophy and achlorhydria. Maintenance of gastric acid secretion may prevent cancer development.

- Mucosal atrophy leads to gastric intestinal metaplasia, dysplasia, and eventually adenocarcinoma.
- Mucosa-associated lymphoid tissue tumors (MALT lymphomas) are also associated with *H. pylori* infection. They are non-Hodgkin B-cell lymphomas arising from the marginal zone of lymphoid follicles.

Ulcerative colitis and colorectal cancer

- Inflammatory bowel disease (ulcerative colitis and Crohn's disease) is the result of an altered host response to intestinal flora, leading to recurrent inflammation. It is a well-known risk factor for colorectal cancer development.
- Risk of colorectal cancer in ulcerative colitis is associated with long disease duration, earlier age of onset, pancolitis and concomitant primary sclerosing cholangitis.
- The carcinogenesis pathway in ulcerative colitis differs from the adenoma-carcinoma sequence encountered in sporadic colorectal cancer.
- Chromosomal instability is detected in nondysplastic tissue in ulcerative colitis patients with long-standing pancolitis.
- Inflammation leads to ROS production, telomere fusion, chromosomal end-fusion, with resulting accumulation of chromosomal alterations.
- The initiating genetic mutations in ulcerative colitis carcinogenesis are usually associated with p53 inactivation and not APC mutations, which occur in sporadic colorectal cancer.
- Inflammation leads to alteration in epigenetic events, such as DNA methylation, transcriptional silencing, inappropriate activation/inactivation of histone deacetylase (HDAC) and acetyltransferases (HAT). This ultimately leads to dysplasia.
- "Field cancerization" has been identified in ulcerative colitis patients. This refers to a clonally expanded area (patch) of mutated cells which share the same founder mutation. The tumor and nondysplastic mucosa also share this same founder mutation.
- The sequence of nondysplastic colitis mucosa to carcinoma is characterized by early p53 mutation, epigenetic silencing, with KRAS and APC mutations occurring late rather than earlier.

Inflammation and cholangiocarcinoma

- Risk factors associated with cholangiocarcinoma include primary sclerosing cholangitis in Western countries, congenital lesions such as choledochal cysts, parasitic infections with *Clonorchis sinensis* and *Opisthorchis viverrini* in Asian countries, as well as hepatitis B/C and *Salmonella* infection.
- Chronic cholangitis resulting from the conditions mentioned earlier leads to inflammation. This leads to release of the proinflammatory cytokine IL-6, a key signaling cytokine in cholangiocarcinoma (CCA) carcinogenesis. IL-6 leads to activation of mitogen activated protein kinases (MAPK), and CCA cell proliferation. Additionally, IL-6 upregulates Mcl-1, an antiapoptotic molecule.
- Inflammation also leads to radical NOS production, DNA damage, and mutagenesis. Chronic exposure of bile duct cells to bile salts leads to persistent activation of EGFR and CCA proliferation. EGFR phosphorylation also leads to MAPK activation, COX-2 production in CCA cells, CCA proliferation, and inhibition of apoptosis.
- Growth of CCA cells is affected by estrogens, neuropeptides, neuroendocrine hormones, and 17-β-estradiol.

Inflammation and hepatocellular carcinoma

- Most cases of hepatocellular carcinoma (HCC) develop in cirrhotic patients. In Western countries, these tumors develop in cirrhosis secondary to alcohol abuse and hepatitis C; in Asia, cirrhosis secondary to hepatitis B and C (HBV, HCV) is the most common cause.
- Hepatic cellular infection with either HBV or HCV leads to increased hepatic production of Hh ligands. Thus, hepatic progenitor cells are activated, leading to proliferation. Additionally, hepatic stellate cells synthesize collagen, ultimately leading to cirrhosis.
- Hepatic cell death in cirrhosis induces inflammation, oxidative stress, and ultimately mutagenesis. Chemokine and cytokine production also foster cellular proliferation and lead to inhibition of apoptosis.
- HBV viral DNA may integrate into host genome leading to loss of genetic stability and chromosomal loss. HCV, an RNA virus, does not integrate into the host genome, but activates NF-κβ, ultimately leading to proliferation of HCC.

- Nonalcoholic steatohepatitis (NASH) also may progress to HCC. Chronic increased dietary intake of fat stimulates production of TNFα, IL-6, stimulation of macrophages and polymorphonucleocytes, thus inducing hepatic steatosis and chronic hepatic inflammation. Chronic oxidative stress resting from this process leads to hepatocyte damage, mutagenesis, and HCC.

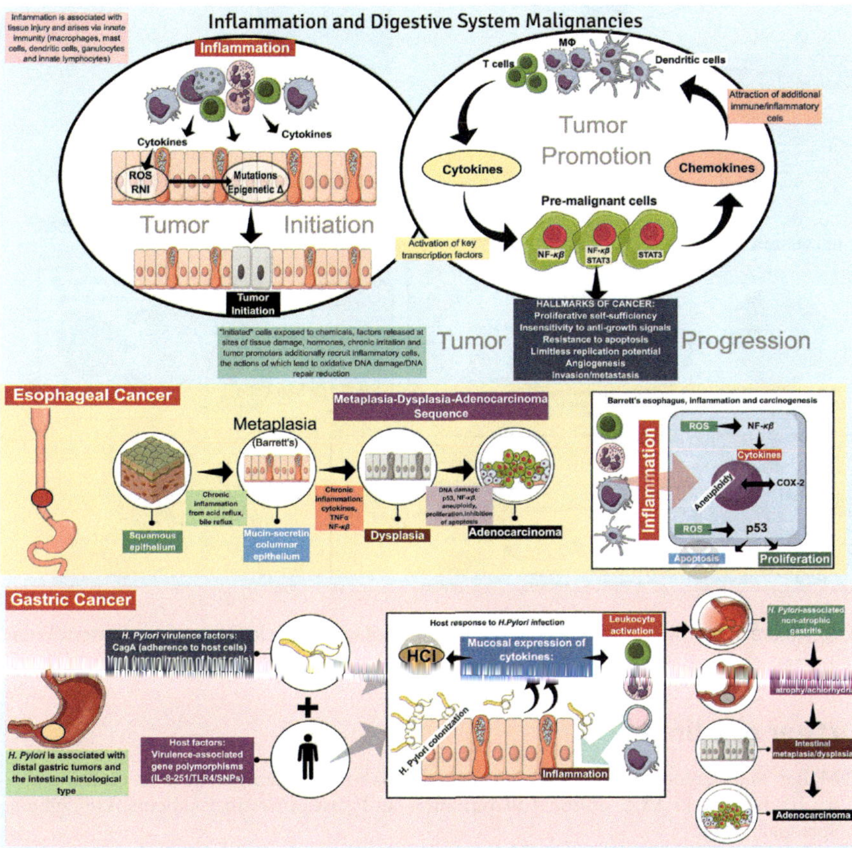

Fig. 1 Tumor initiation, tumor progression: Esophageal and gastric cancer.

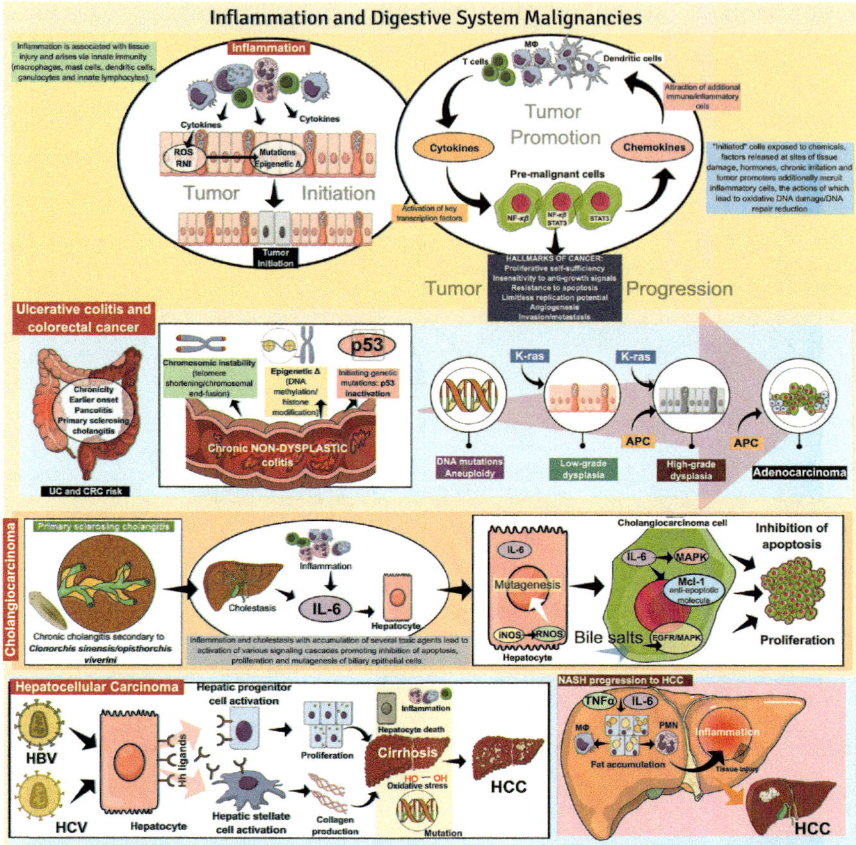

Fig. 2 Ulcerative colitis and colorectal cancer, cholangiocarcinoma and hepatocellular carcinoma.

Further reading

[1] Maher SG, Reynolds JV. Basic concepts of inflammation and its role in carcinogenesis. In: Jankowski JAZ, editor. Inflammation and gastrointestinal cancers. Heidelberg: Springer; 2011. p. 1–24.

[2] Nicholson A, Jankowski J. Acid reflux and oesophageal cancer. In: Jankowski JAZ, editor. Inflammation and gastrointestinal cancers. Heidelberg: Springer; 2011. p. 65–79.

[3] Rathbone M, Rathbone B. *Helicobacter pylori* and gastric cancer. In: Jankowski JAZ, editor. Inflammation and gastrointestinal cancers. Heidelberg: Springer; 2011. p. 83–93.

[4] Jawad N, Direkze N, Leedham SJ. Inflammatory bowel disease and colon cancer. In: Jankowski JAZ, editor. Inflammation and gastrointestinal cancers. Heidelberg: Springer; 2011. p. 99–111.

[5] Yang YM, Kim SY, Seki E. Inflammation and liver cancer. Semin Liver Dis 2019;39(1):26–42.

[6] Coussens LM, Werb Z. Inflammation and cancer. Nature 2002;420(6917):860–7.

[7] Miliewicz P, Wunsch E. Primary sclerosing cholangitis. In: Jankowski JAZ, editor. Inflammation and gastrointestinal cancers. Heidelberg: Springer; 2011. p. 117–28.

[8] Alison MR, Nicholson Linda J, Lin W. Chronic inflammation and hepatocellular carcinoma. In: Jankowski JAZ, editor. Inflammation and gastrointestinal cancers. Heidelberg: Springer; 2011. p. 136–46.

CHAPTER 20

Chemokines, immunity, and cancer

- The tumor microenvironment is the primary location in which tumor cells and the host immune system interact. Different host immune cell subsets are recruited into the tumor microenvironment via interactions between chemokines and chemokine receptors.
- These interactions have a distinct effect on tumor progression or regression, with implications on therapeutic outcomes.
- Chemokines are small proteins which mediate immune cell trafficking and lymphoid tissue development. There are four main classes of chemokines (CC, C, CX3C, and CXC).
- Chemokines are expressed by tumor cells, immune cells, and stromal cells. Their effect is to make immune cells migrate into the tumor microenvironment and regulate tumor immune responses in a spatiotemporal manner.
- Chemokines may target nonimmune cells in the tumor microenvironment, such as tumor cells and vascular-endothelial cells.
- Essentially, chemokines affect tumor immunity, shape tumor immune and biological phenotypes and influence cancer progression, regression, therapy, and outcomes.
- Immune cells with antitumor effects may be recruited by chemokine-chemokine receptor pathways. Examples of such cells are:
  - T-H(helper)17 cells
  - T-H1 cells
  - CD8 + T cells
  - NK (natural killer) cells
- These cells are effector immune cells which may interact with antigen presenting cells, ultimately leading to tumor regression.

Digestive System Malignancies
https://doi.org/10.1016/B978-0-323-98369-3.00002-2

- Pro-tumor chemokine effects may also occur. These chemokines recruit pro-cancer immune cells into the tumor microenvironment. Examples of such cells include:
  - Granulocytes
  - Monocytes (myeloid-derived suppressor cells/MDSC's)
  - Regulatory cells (Treg, T-H22 cells)
  - Plasmacytoid dendritic cells
- In addition, these cells may inhibit antitumor responses, promoting angiogenesis and cell "stemness," thus contributing to cancer progression.
- Tumors may have a immunologically "cold" or "hot" phenotype which reflects on their respective biological behavior.
- A phenotypically "cold" tumor is immunologically enriched in immunosuppressive cytokines which recruit high numbers of pro-tumor immune cells into the tumor microenvironment. Antitumor immune cells are fewer in this microenvironment. Biologically the tumor cells are stem cell-like (ability to evolve into a variety of other tumor cells; less-differentiated cells) and not responsive to immunotherapy.
- A phenotypically "hot" tumor is immunologically enriched in antitumor chemokines which recruit high numbers of antitumor immune cells (effector immune cells) into the tumor microenvironment. Biologically the tumor cells are highly differentiated and may be responsive to immunotherapy.
- Based on current findings it is predicted that directly targeting both pro-tumor and antitumor chemokine—chemokine receptor signaling in combination with other immunotherapies could achieve clinical benefits in patients with cancer.

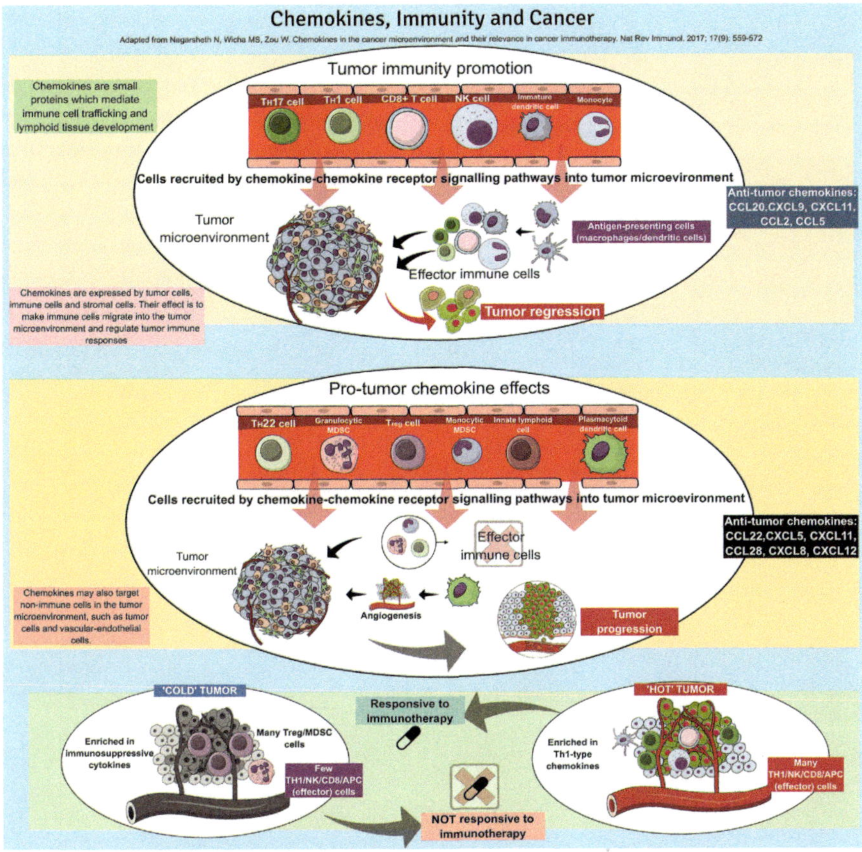

Fig. 1 Chemokines, immunity, and cancer.

Further reading

[1] Nagarsheth N, Wicha MS, Zou W. Chemokines in the cancer microenvironment and their relevance in cancer immunotherapy. Nat Rev Immunol 2017;17(9):559–72.

[2] Zou W, Wolchok JD, Chen L. PD-L1 (B7-H1) and PD-1 pathway blockade for cancer therapy: mechanisms, response biomarkers, and combinations. Sci Transl Med 2016;8. 328rv324.

[3] Zou W. Immunosuppressive networks in the tumour environment and their therapeutic relevance. Nat Rev Cancer 2005;5:263–74.

Captions

Chapter 1. (Esophageal cancer)

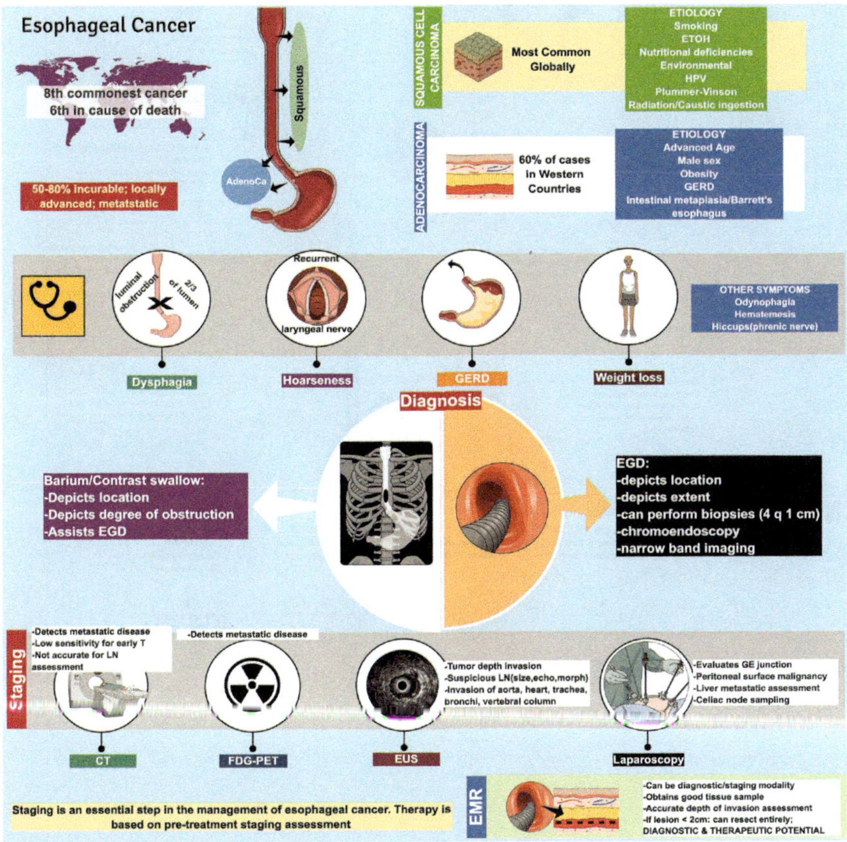

Fig. 1 Esophageal cancer: Epidemiology and evaluation.

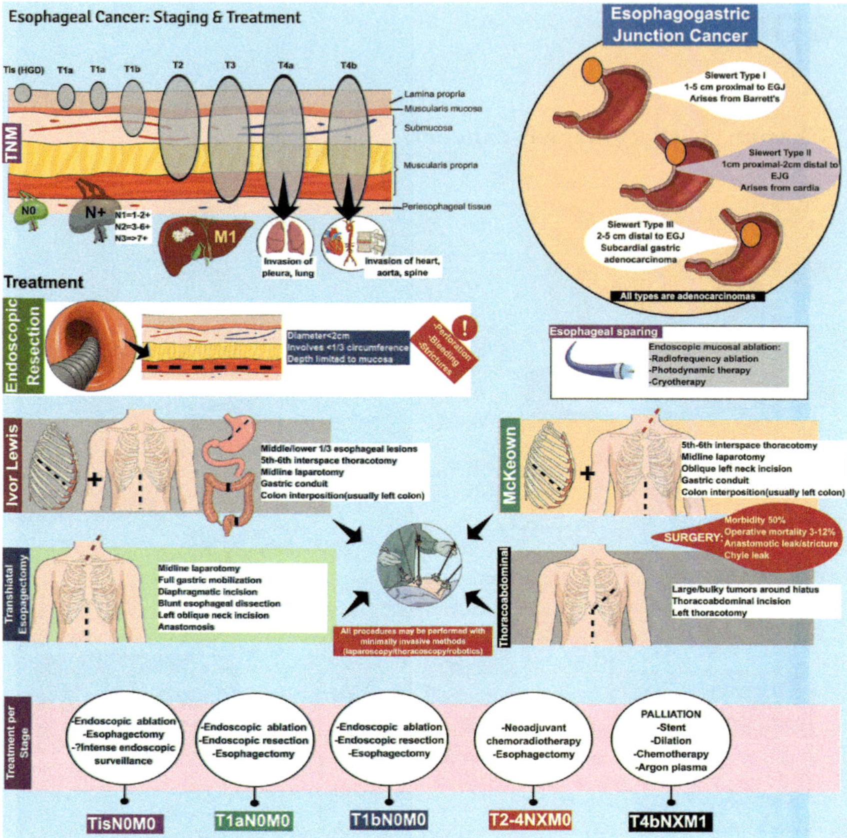

Fig. 2 Esophageal cancer: Staging and treatment.

Chapter 2. (Gastric cancer)

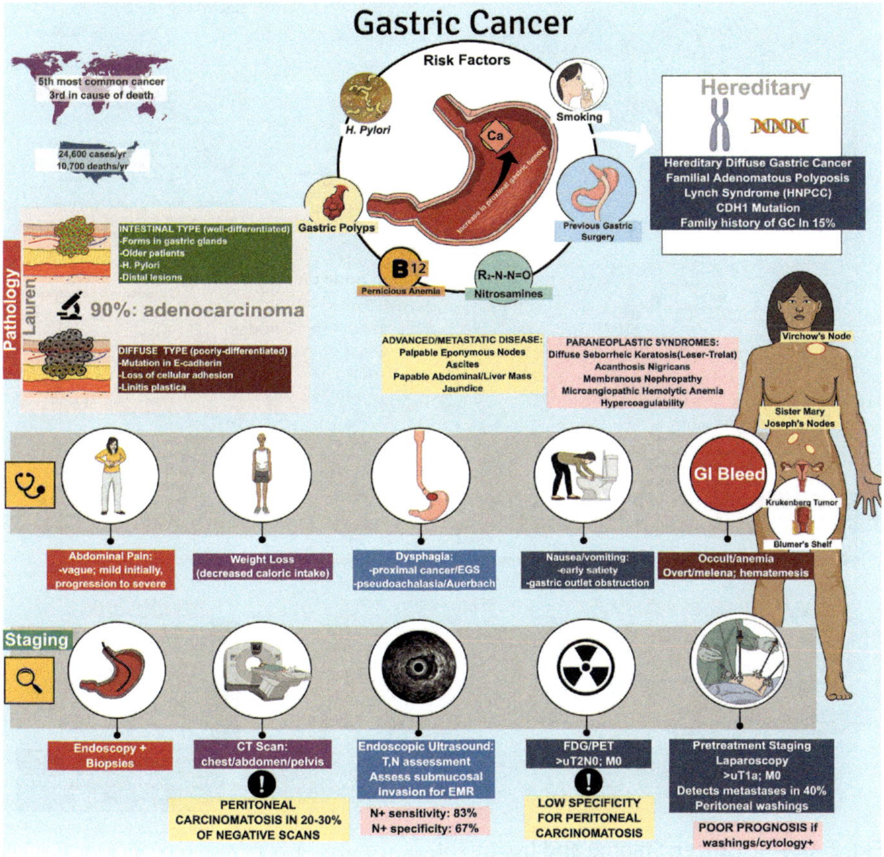

Fig. 1 Gastric cancer: Risk factors, classification, and evaluation.

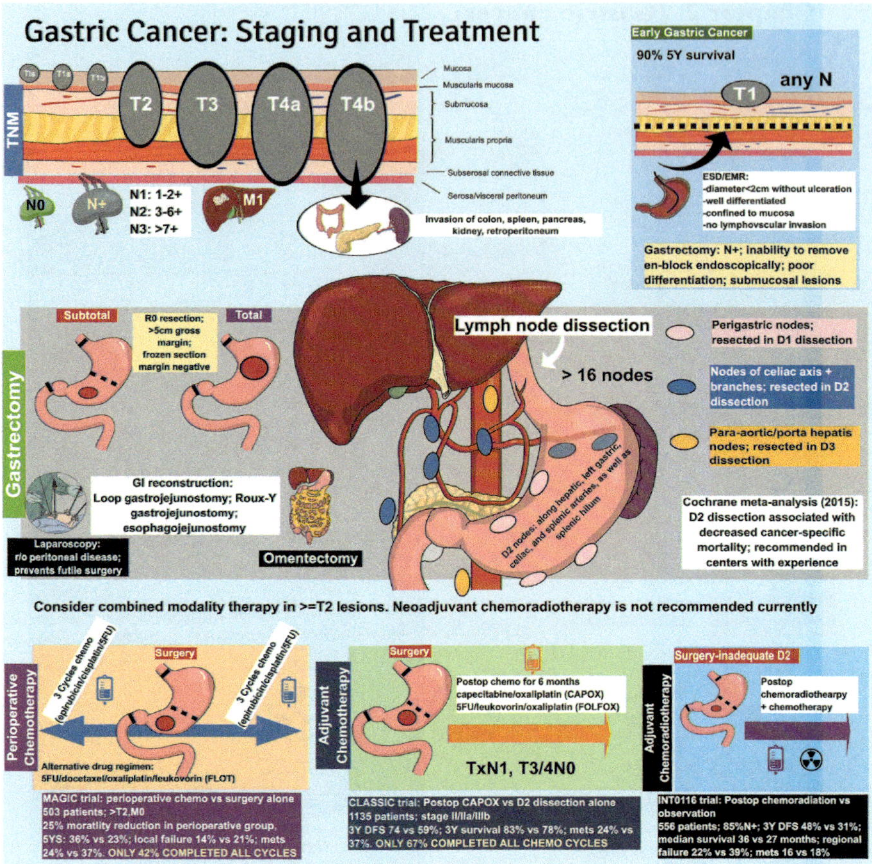

Fig. 2 Gastric cancer: Staging and treatment.

Chapter 3. (Primary gastrointestinal lymphoma)

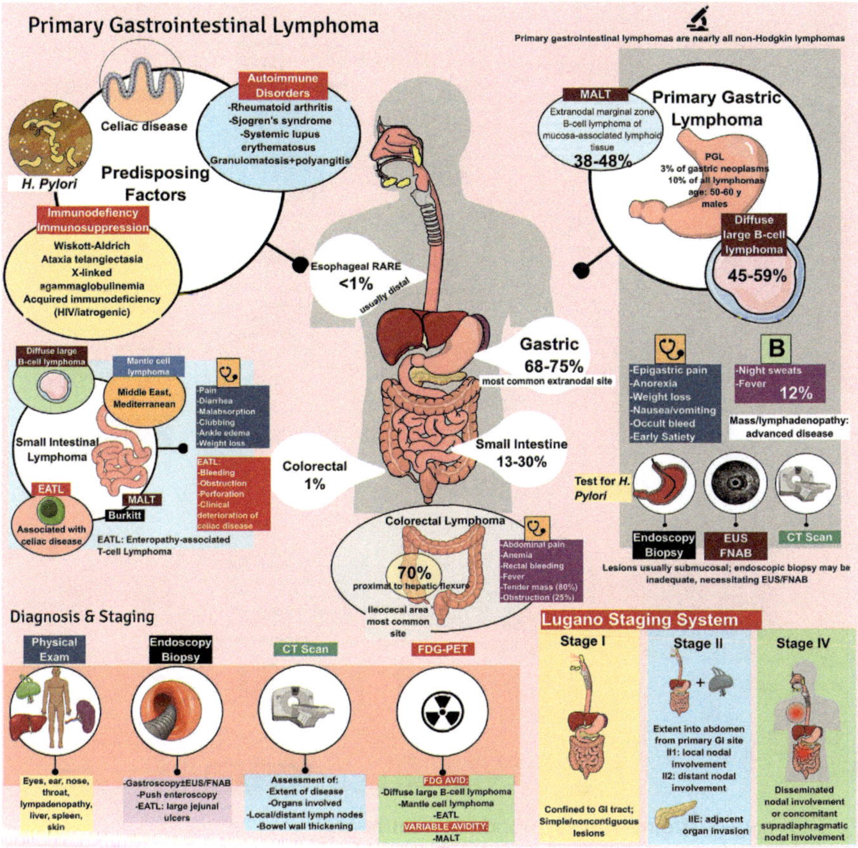

Fig. 1 Primary gastrointestinal lymphoma: Classification, diagnosis, and staging.

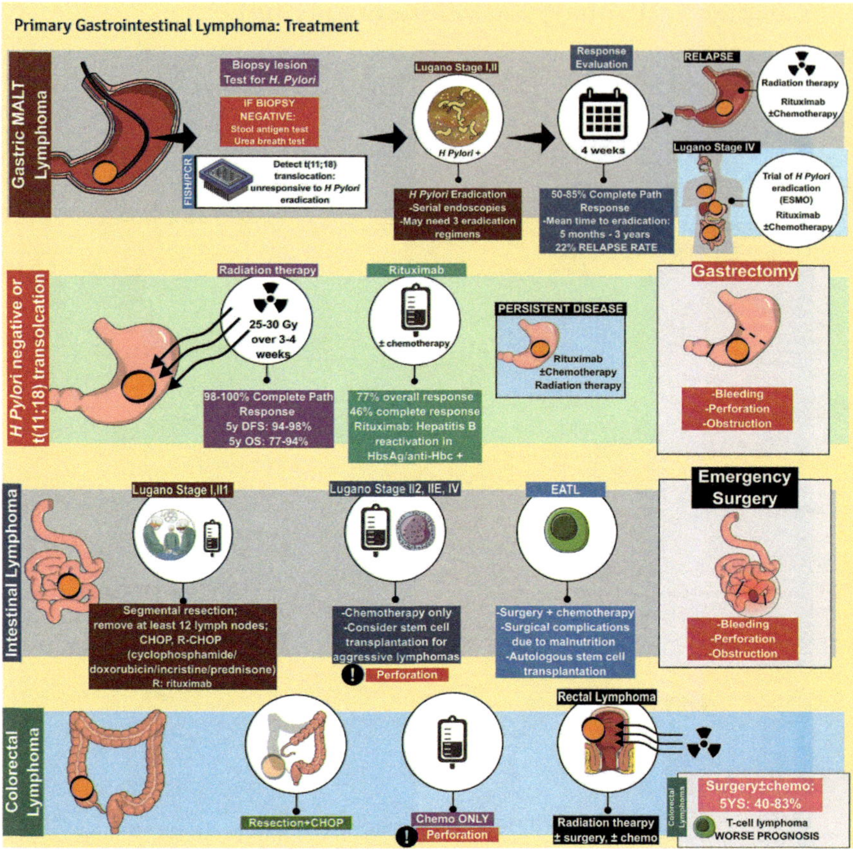

Fig. 2 Primary gastrointestinal lymphoma: Treatment.

Chapter 4. (Gastrointestinal stromal tumors (GIST))

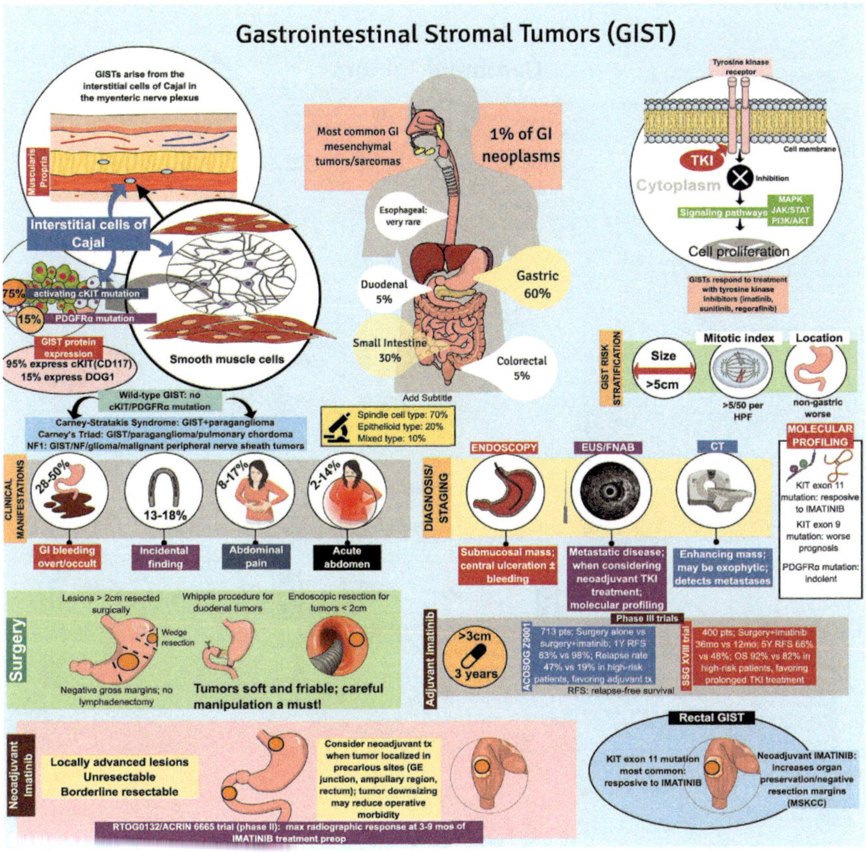

Fig. 1 Gastrointestinal stromal tumors.

Chapter 5. (Gastrointestinal neuroendocrine (carcinoid) tumors)

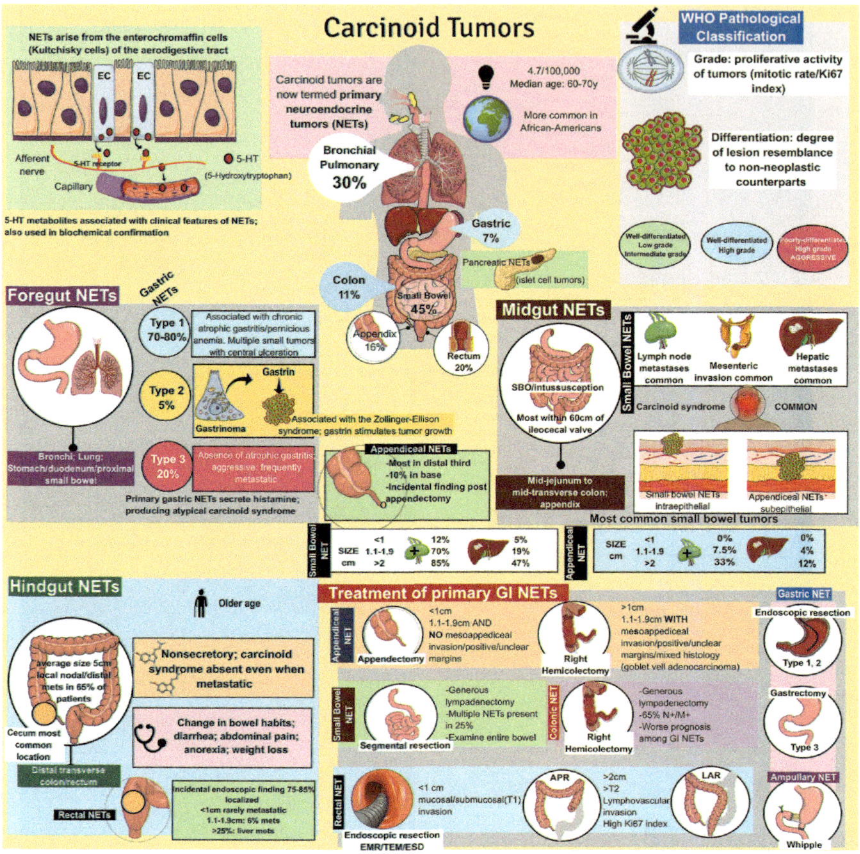

Fig. 1 Neuroendocrine tumors: Classification and treatment.

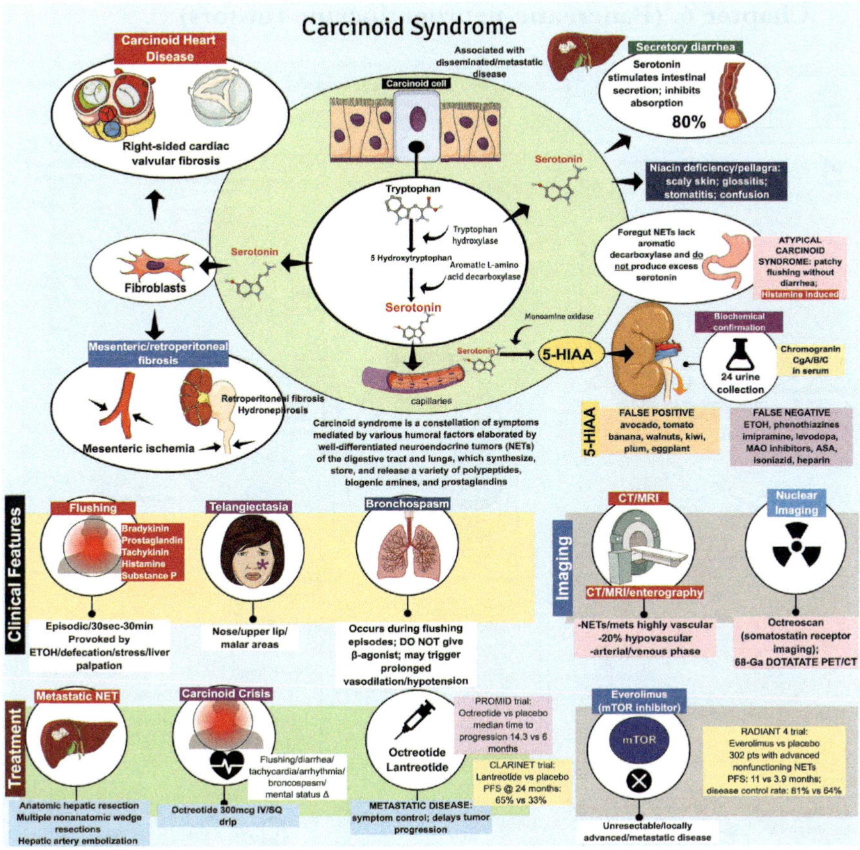

Fig. 2 Carcinoid syndrome.

Chapter 6. (Pancreatic neuroendocrine tumors)

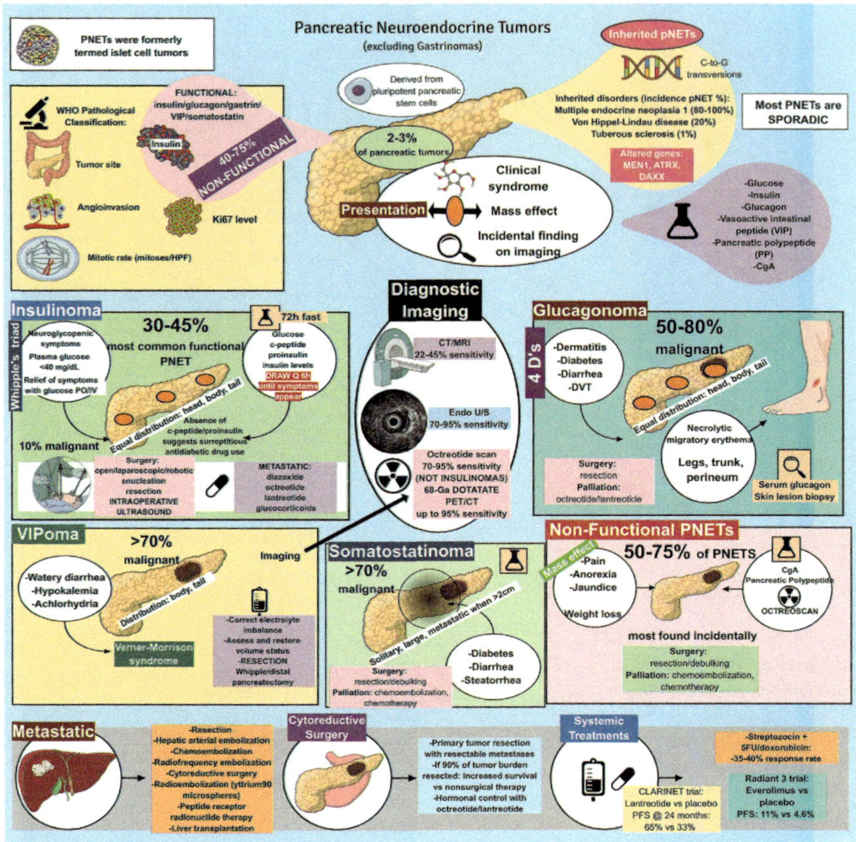

Fig. 1 Pancreatic neuroendocrine tumors.

Chapter 7. (Gastrinoma (Zollinger-Ellison syndrome))

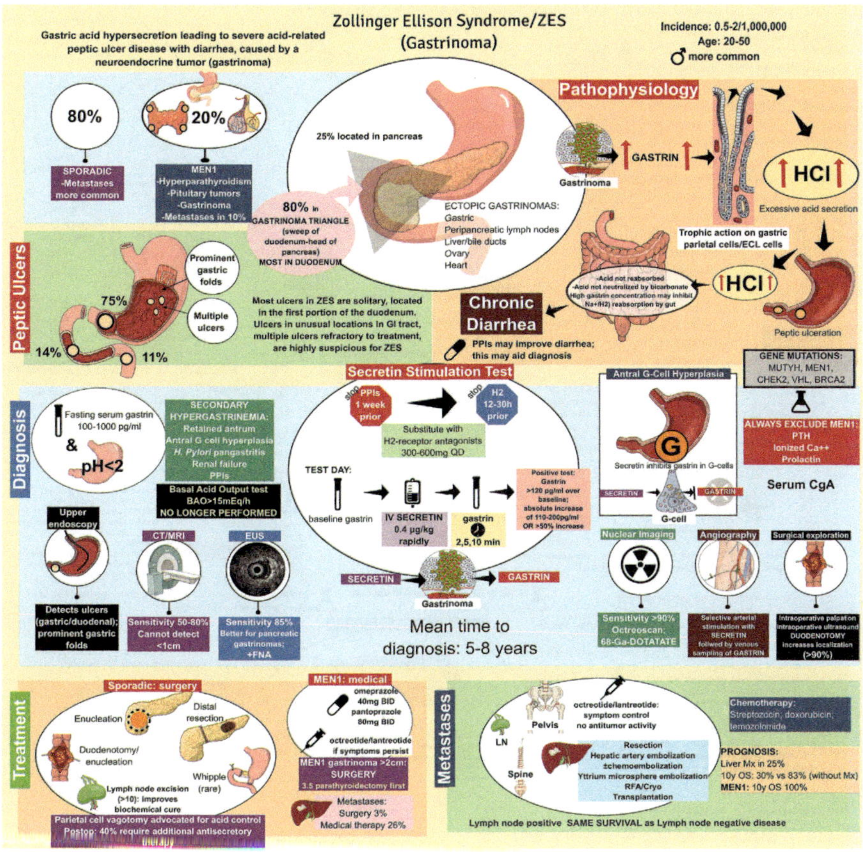

Fig. 1 Gastrinoma.

Chapter 8. (Pancreatic adenocarcinoma)

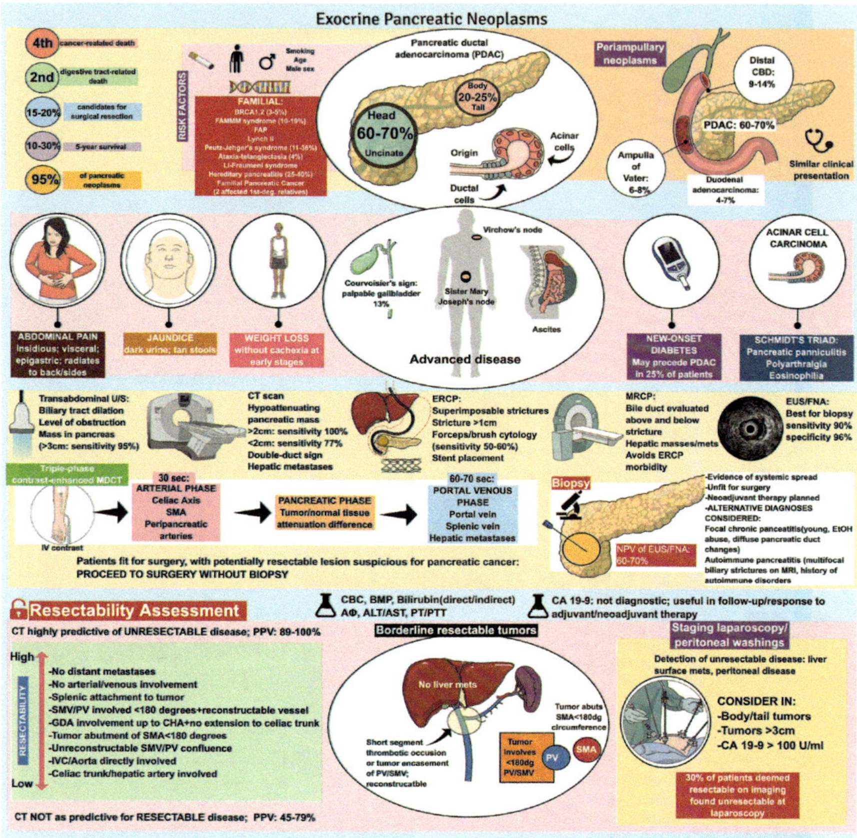

Fig. 1 Exocrine pancreatic neoplasms: Epidemiology and evaluation.

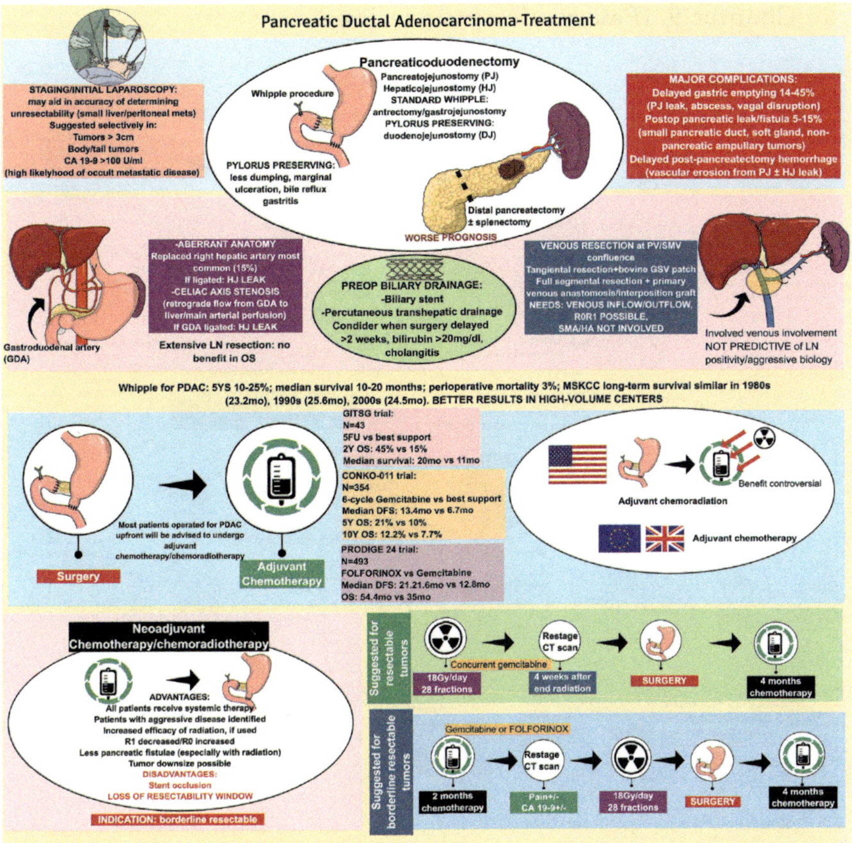

Fig. 2 Exocrine pancreatic neoplasms: Treatment.

Chapter 9. (Pancreatic cystic lesions)

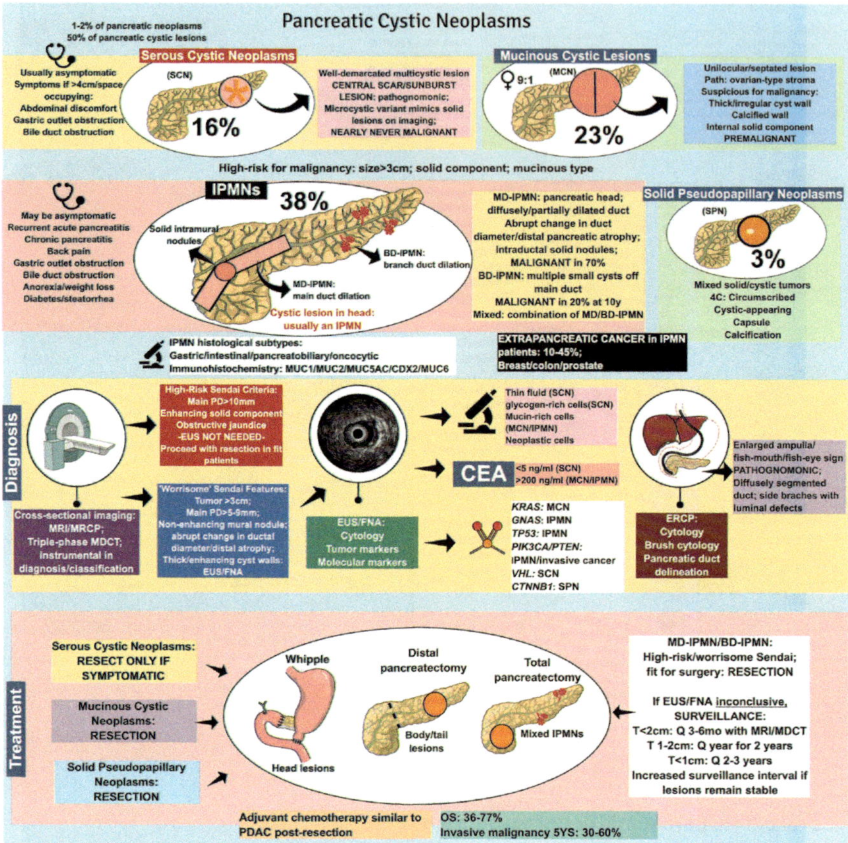

Fig. 1 Pancreatic cystic neoplasms: Classification, diagnosis, and treatment.

Chapter 10. (Hepatocellular carcinoma)

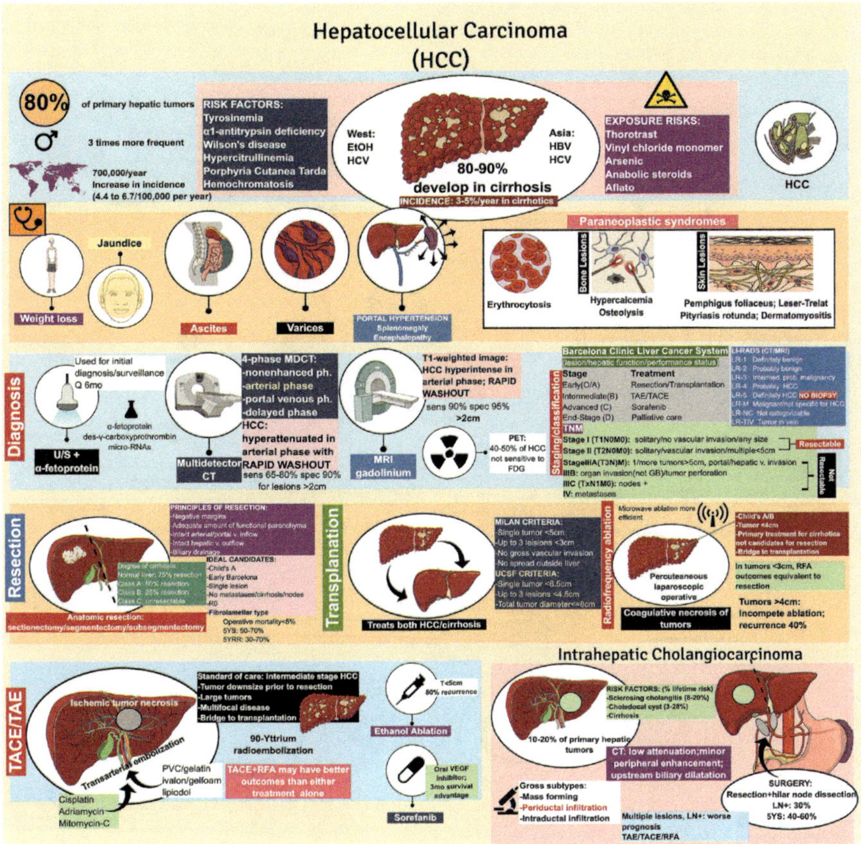

Fig. 1 Hepatocellular carcinoma.

Chapter 11. (Metastatic colorectal carcinoma (CRC))

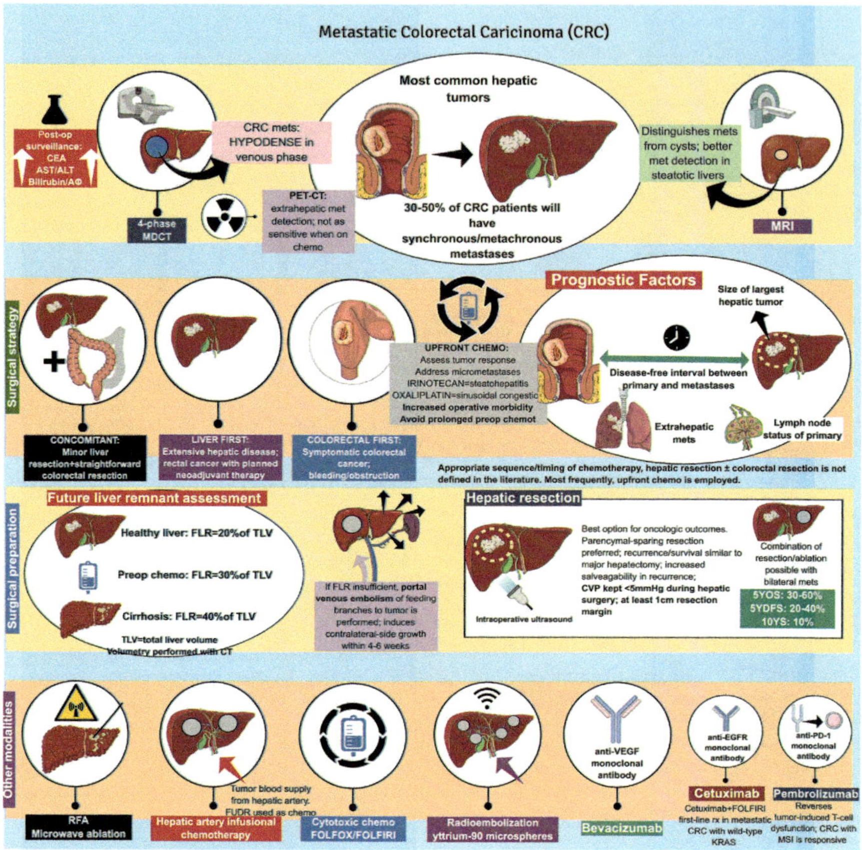

Fig. 1 Metastatic colorectal carcinoma.

Chapter 12. (Hilar, bile duct cancer, and cholangiocarcinoma)

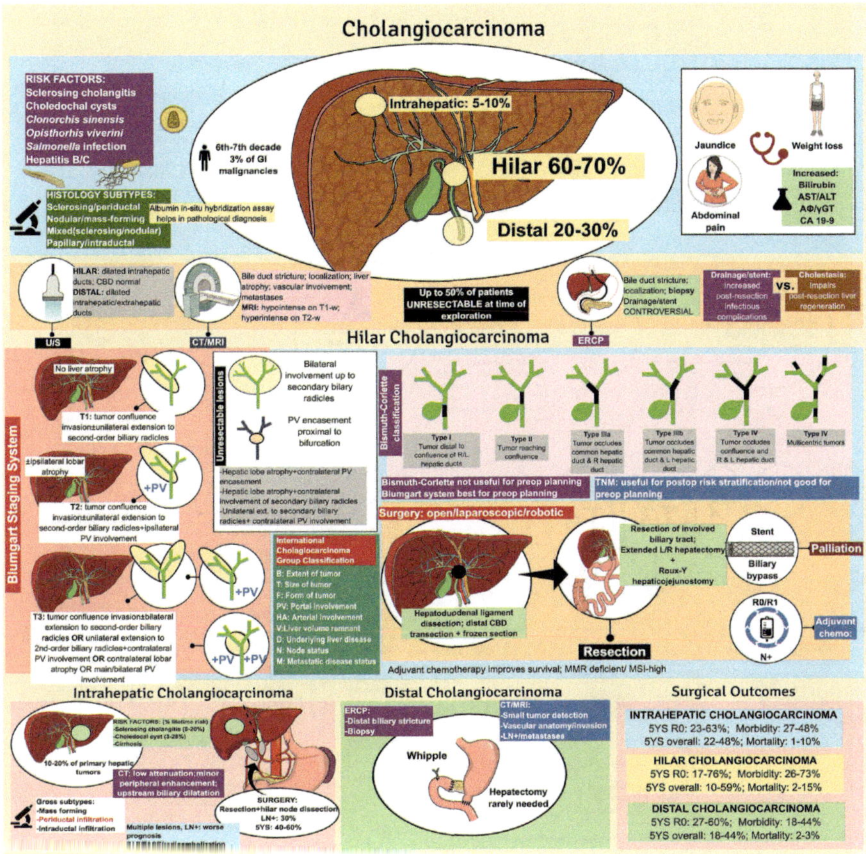

Fig. 1 Hilar and distal bile duct carcinoma: Classification, staging, and treatment.

Chapter 13. (Gallbladder cancer)

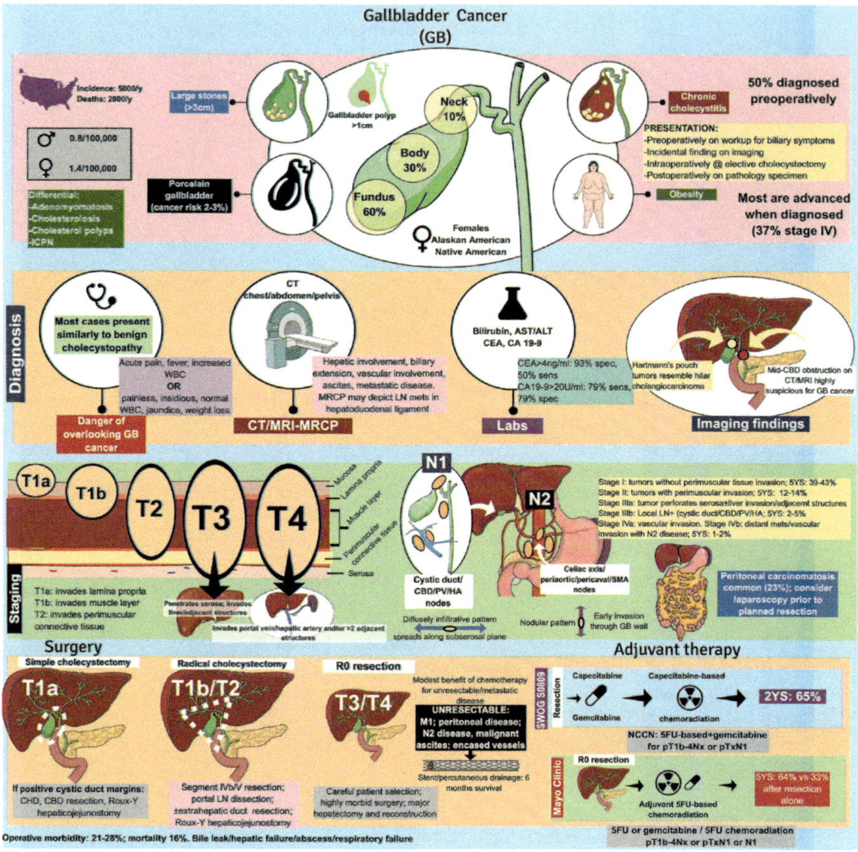

Fig. 1 Gallbladder cancer.

Chapter 14. (Appendiceal neoplasms)

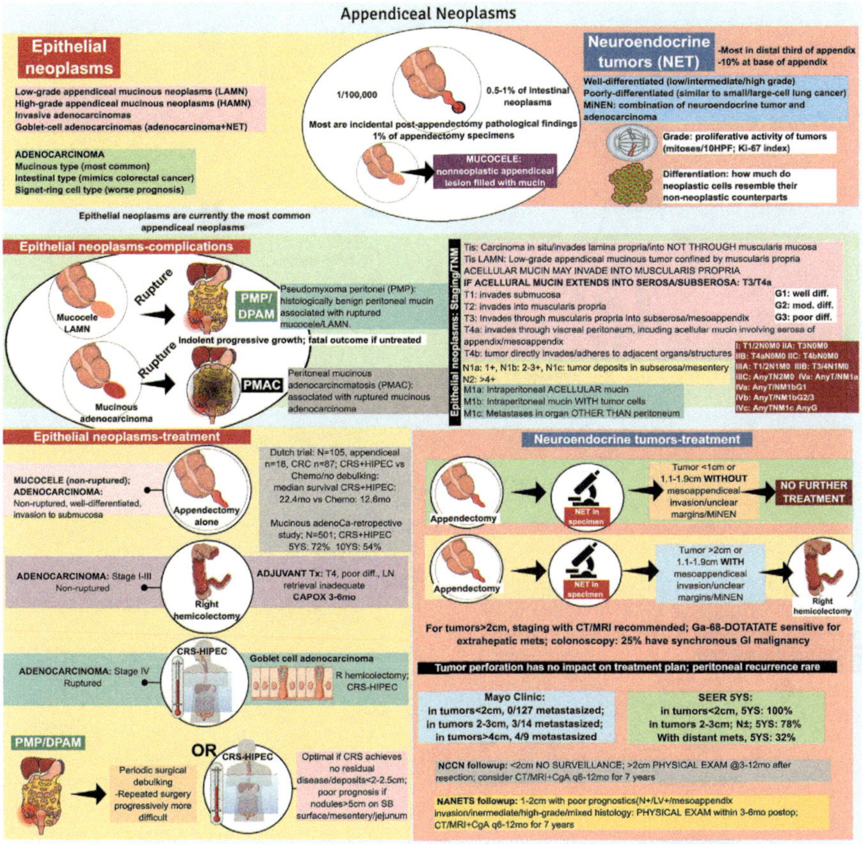

Fig. 1 Appendiceal neoplasms.

Chapter 15. (Colon cancer)

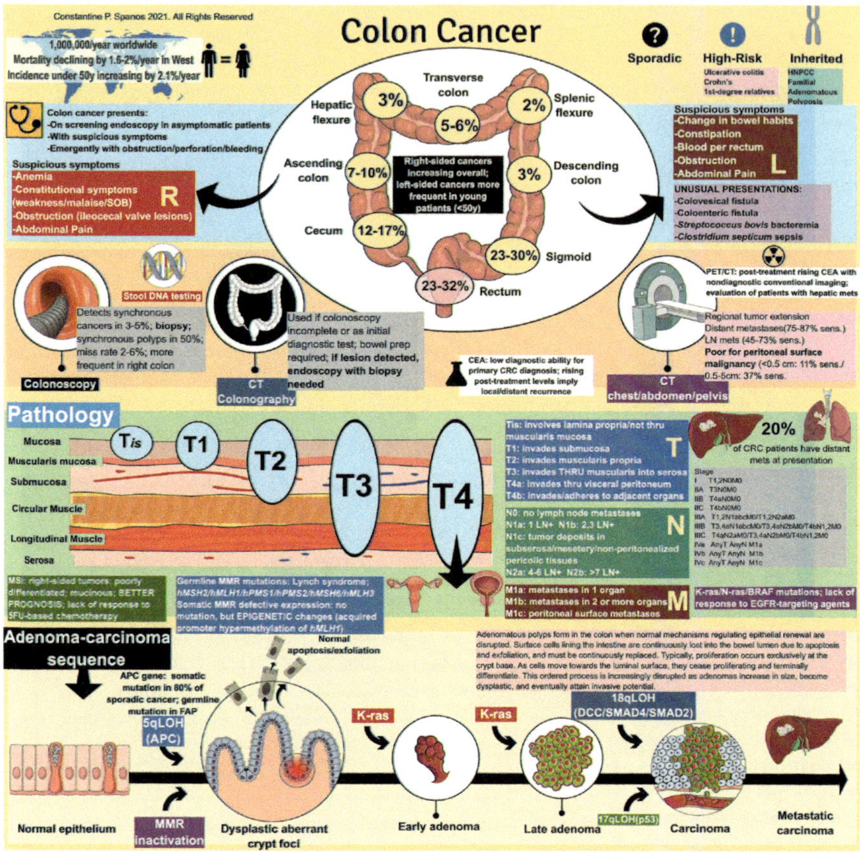

Fig. 1 Colon cancer: Classification, evaluation, adenoma-carcinoma sequence.

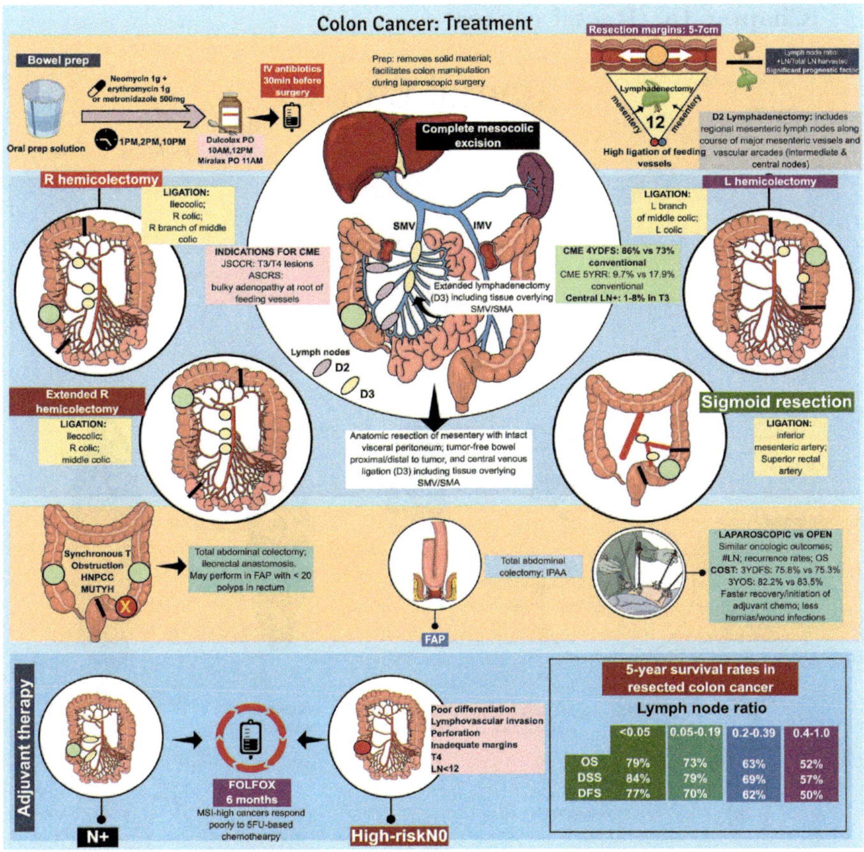

Fig. 2 Colon cancer: Treatment.

Chapter 16. (Rectal cancer)

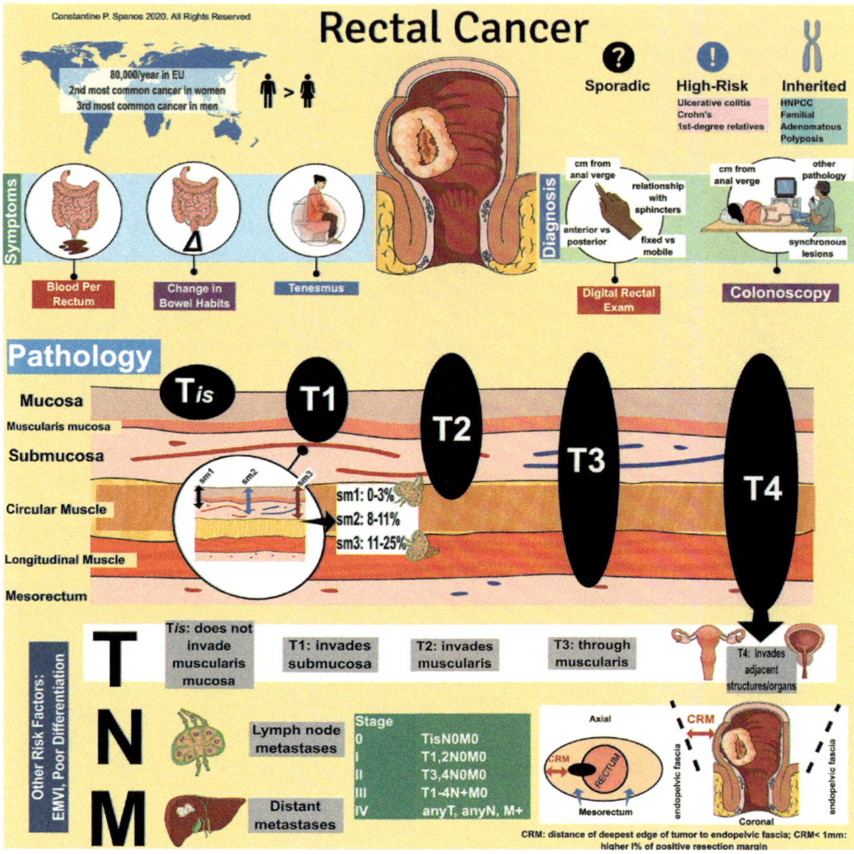

Fig. 1 Rectal cancer: Symptoms and staging.

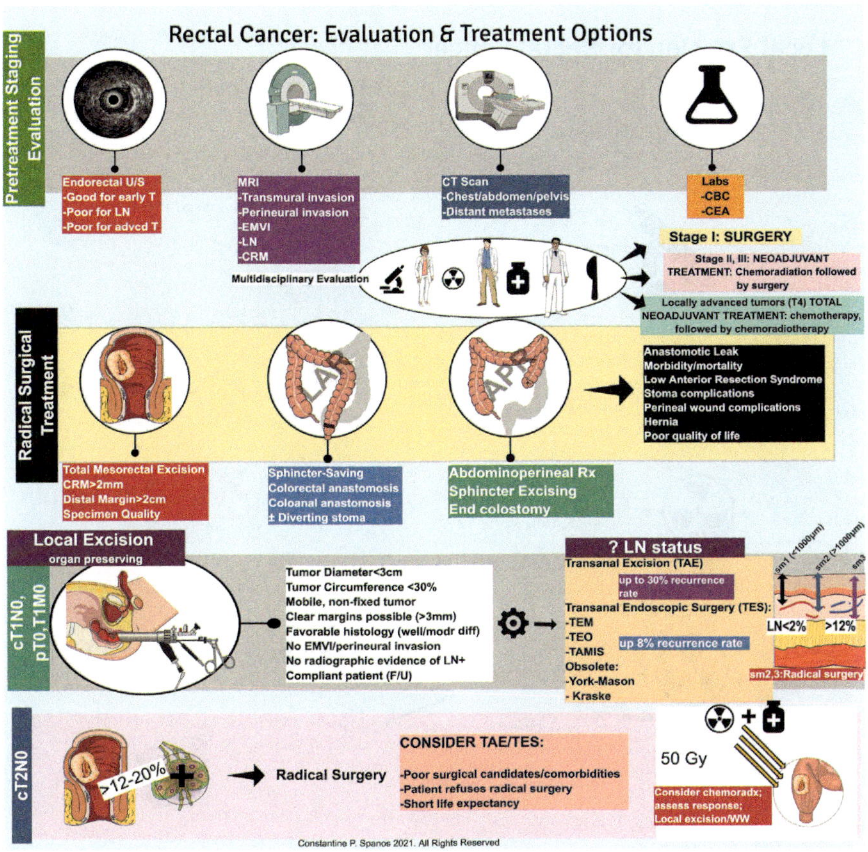

Fig. 2 Rectal cancer: Evaluation and treatment options.

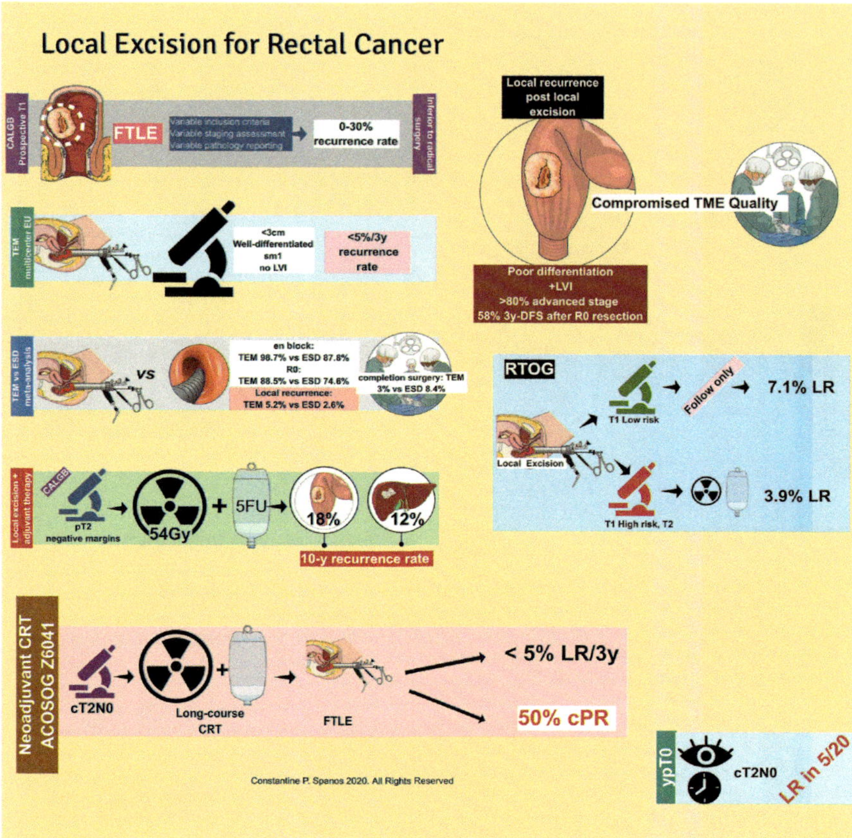

Fig. 3 Rectal cancer: Local excision for rectal cancer.

Chapter 17. (Anal cancer)

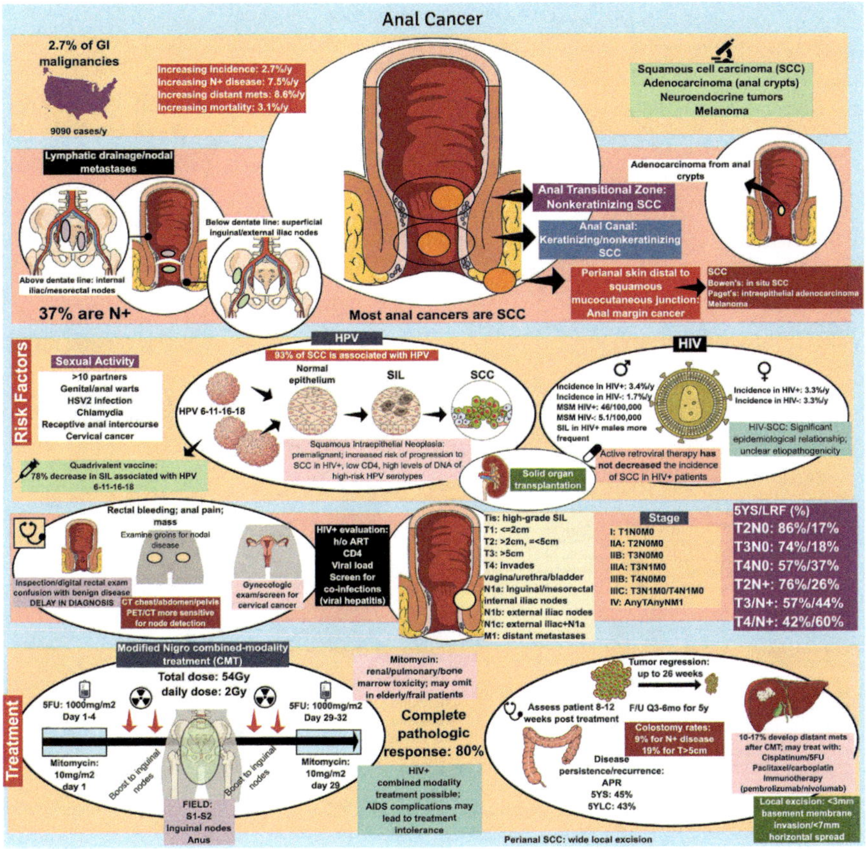

Fig. 1 Anal cancer.

Chapter 18. (Peritoneal surface malignancy)

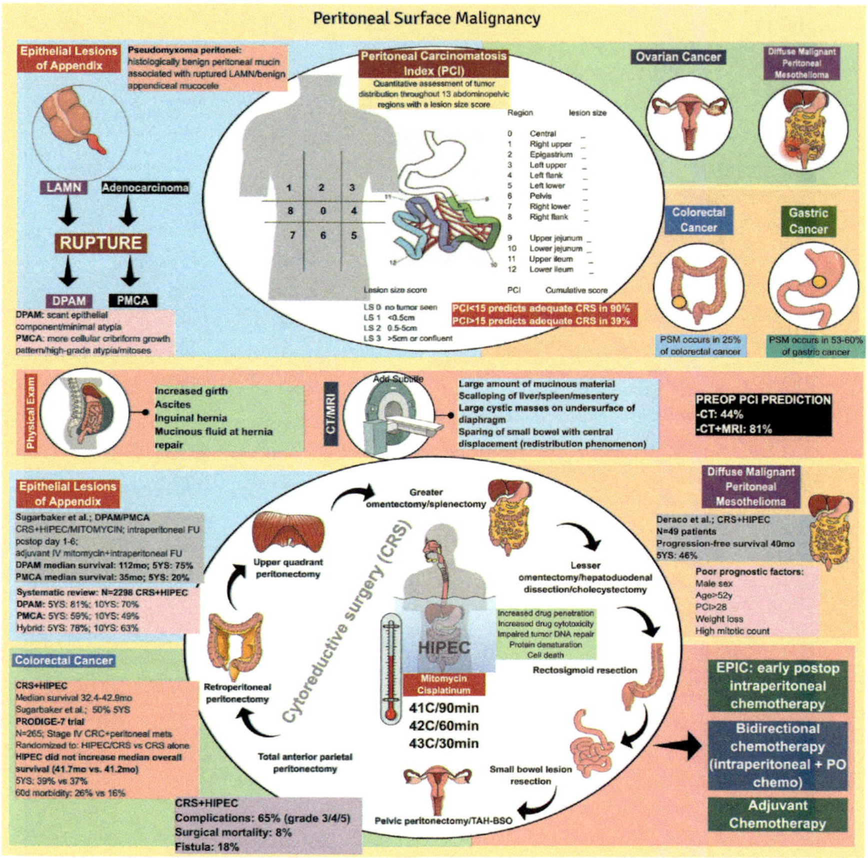

Fig. 1 Peritoneal surface malignancy.

Chapter 19. (Inflammation and digestive malignancies)

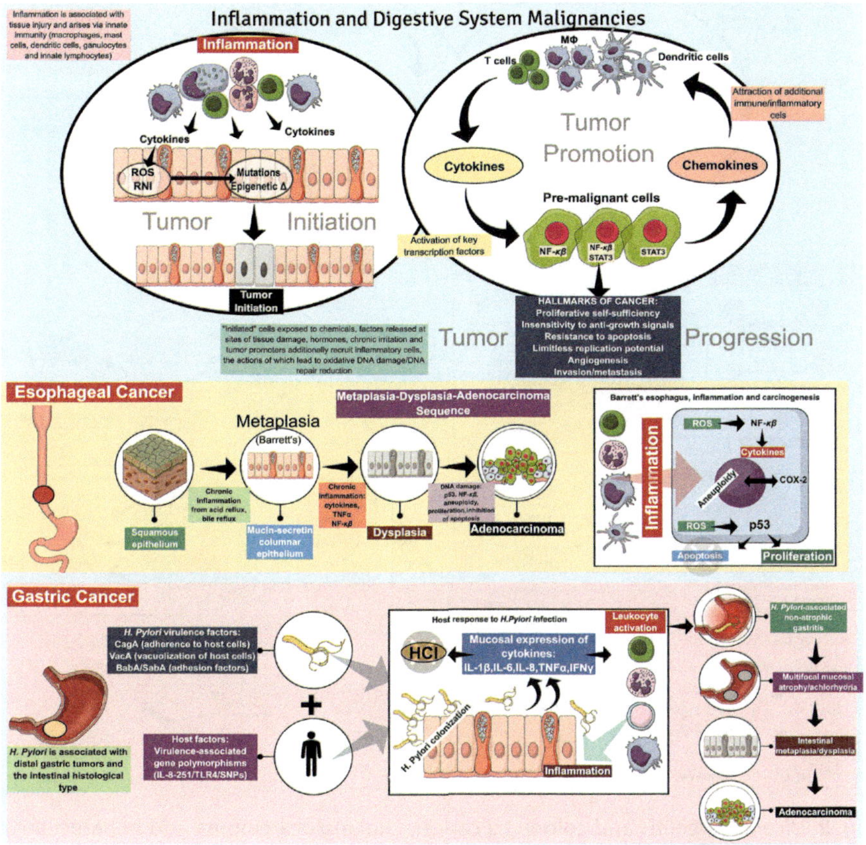

Fig. 1 Tumor initiation, tumor progression: Esophageal and gastric cancer.

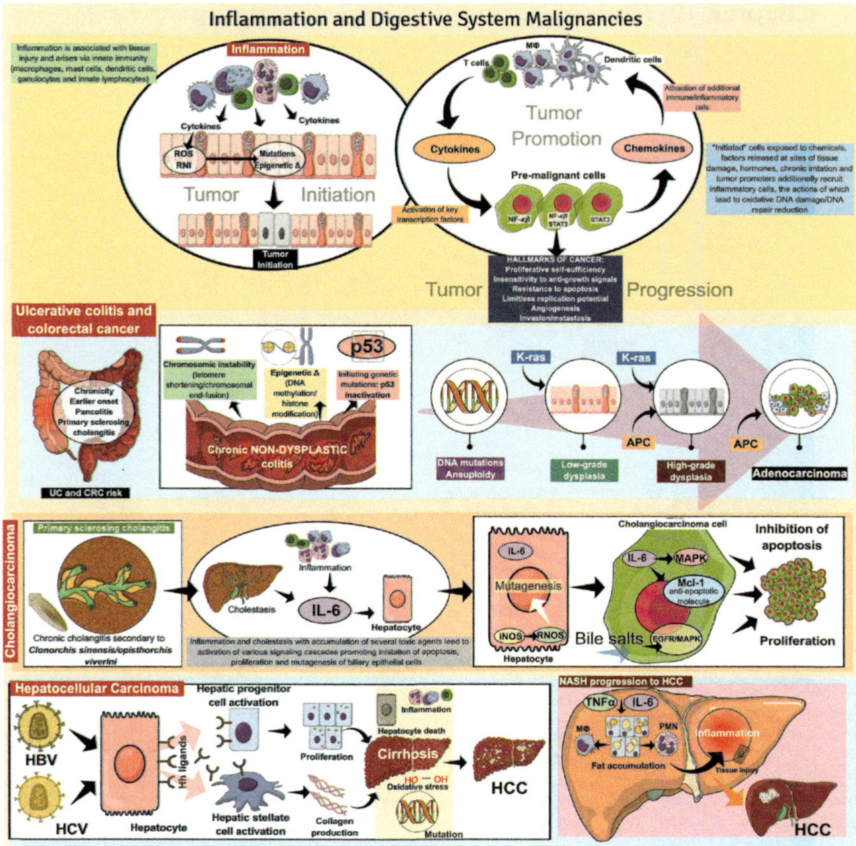

Fig. 2 Ulcerative colitis and colorectal cancer, cholangiocarcinoma and hepatocellular carcinoma.

Chapter 20. (Chemokines, immunity, and cancer)

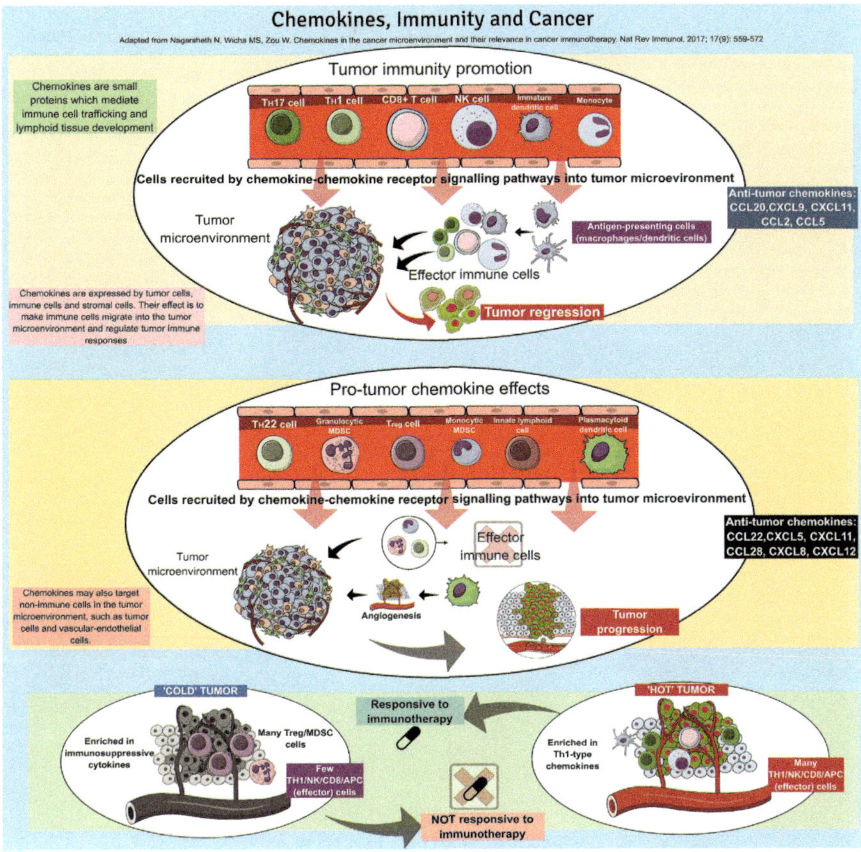

Fig. 1 Chemokines, immunity, and cancer.

Index

Note: Page numbers followed by *f* indicate figures.

9780323983693